Haddington Ventures, L.L.C.

2603 Augusta, Suite 900 • Houston, Texas 77057

COMPLIMENTS OF

HADDINGTON VENTURES, L.L.C.

The Deniers

The Deniers

The world-renowned scientists who stood up against global warming hysteria, political persecution, and fraud*

And those who are too fearful to do so

Lawrence Solomon

RICHARD VIGILANTE BOOKS

PUBLISHED BY RICHARD VIGILANTE BOOKS

Copyright © 2008 by Lawrence Solomon

All Rights Reserved

www.richardvigilantebooks.com

RVB with the portrayal of a Labrador retriever in profile is a trademark

of Richard Vigilante Books

Book design by Charles Bork

Library of Congress Control Number: 2008921479

Applicable BISAC Codes:

BIO015000 BIOGRAPHY & AUTOBIOGRAPHY / Science & Technology

POL044000 POLITICAL SCIENCE / Public Policy / Environmental Policy

SCI026000 SCIENCE / Environmental Science

ISBN 978-0-9800763-1-8

PRINTED IN THE UNITED STATES OF AMERICA

10 9 8 7 6 5 4 3 2 1

First Edition

Dedicated to my daughters, Essie and Catharine,
for their interest in all sides of the global warming debate

Contents

CHAPTER ONE

The Deniers

Global warming has become a question for citizens, and not only scientists. Citizens must decide how serious the threat is and what to do about it, which cures make sense, and which might be worse than the disease. Alas, the answers to these questions depend on scientific issues of fierce complexity that few laymen are capable of confronting directly.

So what are we to do?

Al Gore has an answer, and in some ways it is a very sound answer. Mr. Gore says, essentially, that we must rely on 'the argument from authority.' We must accept the word of experts who know directly what we can 'know' only because they tell us. Go to the scientists and ask them. They have the right training and access to the best data. They understand the equations.

And what the scientists say, according to Gore and the UN and an overwhelming consensus of the media, is that "the science is settled." There is no longer any serious doubt that global warming is a grave problem already, that it is rapidly getting worse, that it is caused primarily by human activity, and that it will lead to catastrophe if those activities continue unchecked.

The argument from authority is perfectly respectable, even essential. Every day, in all types of situations, we must act on the best information available to us, even on matters outside our technical expertise. If a respected doctor tells you that your child has a serious medical condition requiring immediate surgery, you will pay heed, especially if you obtain a second opinion that unambiguously backs up the first. On the other hand, if the second opinion provides diametrically opposite advice, and tells you that your child's condition has been misdiagnosed and that an operation entails needless risks, the argument from authority is no longer quite so useful, at least not immediately. You then need to look more deeply into the subject, perhaps seeking more advice from others, perhaps paying special attention to the relative credibility of the contending authorities.

According to Gore and the UN and most of the media, however, that is just what we do *not* need to do in the case of global warming. To them, there is no dispute worthy of the name. All competent authorities, or at least a "consensus" of such authorities, agree with the doomsday view of global warming. There are no dissenters sufficiently credible that we are bound to take them seriously.

Then what of the "deniers" we have all heard about, those holdouts in the global warming debate, complete with Ph.D.s at the end of their names, who refuse to accept the obvious? Gore and company have a ready answer, repeated again and again: pay no attention. These alleged scientist dissenters are either kooks or crooks who take the pay of the oil companies to spew out junk science and confuse the issue. Here's what Mr. Gore says about them: "Fifteen per cent of the people believe the moon landing was staged on some movie lot and a somewhat smaller number still believe the Earth is flat. They all get together on a Saturday night and party with the global warming deniers."[1] *Newsweek,* in a now famous cover story, called these scientists part of "the denial machine,"[2] funded by the energy industry and organized by corrupt right-wing lobbyists.

The very term "deniers" is a deliberate reference to the "Holocaust deniers" who defend the Nazi regime by claiming that Jews and their allies faked the Holocast to slander Hitler. Scott Pelley, of CBS's *60 Minutes,* was asked by CBS Web reporter Brian Montopoli why he "did not pause to acknowledge global warming skeptics"[3] in his influential broadcasts on the topic. Pelley replied, "If I do an interview with Elie Wiesel, am I required as a journalist to find a Holocaust denier?"[4] According to Montopoli, Pelley went on to explain that "his team tried hard to find a respected scientist who contradicted the prevailing opinion in the scientific community, but there was no one out there who fit that description."[5] Pelley declaimed that he was not interested in "pseudo-science or conspiracy theory blogs," but "sound science."[6] Other journalists taking the denier's metaphor to its logical conclusion have suggested that climate-change dissidents be put on trial for crimes against humanity.

When I first heard about the deniers, I did not doubt that either lobby groups or scientists could be bought. I work for an environmental group called Energy Probe, one of Canada's largest and oldest, and have seen this firsthand. We have been an anti-nuclear organization since 1974, when we began opposing Canada's nuclear establishment, and know that industry scientists can twist the truth to suit their paymaster.

At the same time, I also know firsthand that scientists with integrity can hold unconventional and unpopular views, because this was the case in the 1970s and 1980s with a set of scientists who were deniers at the time—the small group of scientists who dissented from the conventional wisdom of the day that nuclear power was safe, clean, and inexhaustible. They were scientists of integrity who stuck to their principles despite the scorn heaped on them at the time—unlike today, nuclear power in the 1970s had almost universal acceptance and almost no one in business, government, or academia would risk ridicule by questioning it.

I am also sensitive to false accusations because Energy Probe has sometimes been at the receiving end of them. The nuclear industry, thinking that the anti-nuclear movement was being funded by its competitor—the oil industry—for decades spread rumors that Energy Probe was a front for oil companies. I thought the accusation laughable, since Energy Probe was also the chief critic of the petroleum industry and had succeeded in stopping Arctic pipelines, oil sands plants, and other petroleum industry megaprojects. Yet with some, the accusations stuck, or at least left nagging doubts. Sometimes I didn't even learn of rumors until years later, because they weren't publicized—they were whispering campaigns designed to undermine confidence. As I learned from industry documents that came out during a court case, in which Energy Probe sued the federal government over the constitutionality of an act protecting GE, Westinghouse, and other nuclear players in the event of a nuclear accident, the rumors could be petty and even, to my mind, farcical. My favorite example of this involves nuclear industry attempts to undermine our anti-nuclear spokesman Norm Rubin, who became a household name in Canada because the media had him on TV and radio daily. Norm cannot only out-duel his opponents on the facts, he is telegenic and witheringly funny in his attacks. Documents that the nuclear industry was spreading claimed that Norm lived the high life and owned a Jaguar. In fact, Norm, who is as frugal as they come, owned a 15-year-old Volvo at the time, although almost no one knew that—he always rode a beat-up old bicycle, even in winter. More commonly, the nuclear industry mocked Norm for being a musician, which was true—he had taught music history at Princeton and the University of Toronto. The industry said nothing of Norm's other degree, in science, from MIT.

So when the rhetoric began to heat up on the global warming skeptics, a part of me suspended judgment on the deniers who were being so roundly criticized. Yes, I expected that the oil industry—or any industry—would try to buy off scientists if they could. Had all scientists who questioned global warming been bought? I wasn't prepared to come to that conclusion without evidence. I might not have sought the evidence were it not for Norm.

Before I tell you about Norm's role in this book, you should know that I had a prior, albeit casual, familiarity with the global warming issue. As you might expect from an environmental group, Energy Probe had long been concerned about global warming. We are primarily advocates of conservation and renewable energy as alternatives to fossil fuels, large hydro dams, and nuclear power. We are also a peace group concerned with stopping the proliferation of nuclear weapons around the world (you may be surprised to learn that Canada helped give both India and Pakistan the nuclear bomb). In the late 1980s, we also were among the very first organizations in Canada to sound the alarm on global warming. In 1992, our Third World Division, Probe International, participated in Rio at the international gathering that gave rise to the Kyoto Protocol. And both Energy Probe and Probe International developed proposals for minimizing the threat of global warming.

This book really began, however, with a bet over a dinner in Toronto's Chinatown almost two years ago. Energy Probe and Probe International had invited some fellow environmentalists from China (Chinese environmentalists—and you think you have a tough job) to come to Toronto for an extended visit with us. We were gathered to celebrate their arrival. The conversation turned to global warming when Norm remarked on the science being settled. In part because I knew Chinese environmentalists aren't exposed to environmental debate, in part because I thought it likely that some credible scientists disagreed, and in part because Norm and I give each other no quarter, I challenged Norm to name three climate-change areas that he felt were settled. Probably expressing more confidence than I had at the time, I told him if he identified the areas of expertise, I would find a credible dissenting scientist in each.

Well, the conversation took off on its own as good conversations do, and Norm never did propose the three areas, despite my prodding him. Nevertheless, I thought it would be fun to see if I could find the scientists whose existence I had so boldly predicted. Besides, I also write a weekly column for the *National Post*. Like any journalist with column inches to fill, if I took the trouble to find these scientists, I certainly was going to get a column or two out of them. This would have the added benefit of forcing a response from Norm, because the way things work at Energy Probe, anything controversial that we write gets vetted first by the colleagues most inclined to disagree. Norm, among others, would edit and approve my first few columns.

So on Nov 28, 2006, I wrote my first "Deniers" column for the *National Post*. To date, I have profiled some three-dozen scientists, all recognized leaders in their fields, many of them actually involved in the official body that oversees most of the world's climate-change research, the United Nation's Intergovernmental Panel on Climate Change. Some have even been involved as lead authors. The "Deniers" columns (I still occasionally write them) got by far a greater response from readers than anything I have ever done. Many of those readers were scientists themselves. Their e-mails and phone calls thanking and encouraging me made me feel—well, thanked and encouraged, so I kept plugging away.

But some things about the series frustrated me: the format of a newspaper column doesn't give me much space to work with and, of course, no footnotes and few graphs. Pretty much all I could do was point to the existence of these eminent scientists and leave it at that. Also after more than a year into the work, I had gained insight into both the scientific issues and the ferocious political and media battles that were being waged, some of which had made me an unexpected participant. Those first columns, I could not help thinking, would have been a lot better had I known then what I know now. And issues (and some very public controversies) had arisen that could be effectively addressed only in a book. So I wrote one.

In the book, as in the columns, I follow a few rules. The most important is that I do not attempt to settle the science myself. Herein you will find scientists who disagree profoundly not only with some of their colleagues who support the doomsayer view but with other scientists profiled in this book. Such disagreement is the very stuff of science. More important, I am a layman trying to understand, and help other laymen to understand, how we should think about the global warming debate. For us, the answer cannot be to settle the science directly. For the most part, the layman must rely on the argument from authority, including a careful sifting of the credibility of the authorities and the relevance of their expertise to their particular claims for which they are advanced as witnesses.

The question of credibility brings me to another rule I imposed on myself: I would not play the numbers game. I would not rely on claims that 14,000 scientists signed one petition saying the planet is toast, or that 14,001 signed another saying global warming is a hoax. There are a lot of scientists in the world. By definition most of them are mediocre. Getting thousands of mediocrities to sign a petition is an impressive work of political organizing; it is not science. No, I was looking for a relative handful of scientists of great eminence, whose credibility (unlike their equations) would be transparent to the lay reader.

I have been asked many times why I titled my series and now this book *The Deniers*, in effect adopting their enemies' terminology. Many of the scientists in this book hate the term and deny it applies to them.

I could give several reasons, but here is the most important. The scientists are not alone in having their credibility on trial in the global warming debate. They are not the only "authorities" in the argument, and not even the most important "authorities." Most laymen, most citizens, owe most of what we think we know about global warming not to science directly, but to science as mediated by the media and by political bodies, especially the UN and our governments. We citizens, trying to discern what to do about global warming, must judge not only the credibility of the scientists but of those who claim to tell us what the scientists say. To that end, as you read through this book, judge for yourself the credibility of those who dismiss these scientists as cranks or crooks, and call them *The Deniers*.

CHAPTER TWO

The Case of the Disappearing Hockey Stick

Dr. Edward Wegman

My colleague, Norm, believed that "the science is settled" for good reason. For though I would discover that skeptics abounded in a camp of their own, the science-is-settled camp is very much the more prominent of the two. It certainly is much better funded and gets far better press than the skeptical camp (with both those disparities growing dramatically of late). And without doubt, the science-is-settled camp includes many eminent scientists with impressive credentials.

The skeptics, on the other hand, are harder to find. Many of them, I would discover, don't want to be found at all and try very hard not to appear to be dissenters. They have no wish to be called names in the press, or to lose their jobs, or to have their funding cut off as many deniers have. I have actually had the children of one denier denounce me for smearing their father's good name. The smear? I pointed out in a column that his work contradicts the doomsayer orthodoxy. Lately, this fear of reprisal or loss of reputation seems to have inspired a sort of ritual obeisance among serious scientists who find themselves refuting some part of the case for impending doom. Time and again such scientists, having just heard themselves proudly announcing a scientific discovery that inadvertently shreds part of the science-is-settled case, immediately follow up with a meek recital to the effect that *of course this does not mean that I dispute the gravity of global warming or that humans are the primary cause thereof.*

At the time I started digging up deniers, I knew none of this. I naively imagined that all reputable scientists would be delighted to have their stories told. Some were. Some weren't, and spoke to me grudgingly. Many simply ignored me, perhaps hoping I would go away. And a few absolutely hated it and have never forgiven me for "outing" them.

How powerful must an orthodoxy be if men who are praised for questioning it try furiously to deny they have done any such thing?

Against this background, Dr. Edward J. Wegman was an easy choice for my first column. No one could describe Wegman as being on the fringes of science. Dr. Wegman is a director at the Center for Computational Statistics at George Mason University, chair of the National Academy of Sciences Committee on Applied and Theoretical Statistics, and board member of the American Statistical Association. Few statisticians in the world have CVs to rival his.

Nor had Wegman been reluctant to join the fray or put his name to his dissenting comments. Plus Wegman's work challenged, and many say demolished, one of the hottest and most publicized claims of the science-is-settled side.

Wegman became involved in the global warming controversy after the Energy and Commerce Committee of the U.S. House of Representatives asked him to assess the "hockey stick graph," the IPCC's poster child in the global warming debate. Based on calculations by Michael Mann at the University of Massachusetts, the graph shows that, for most of the past 1,000 years, temperatures in the Northern Hemisphere have been slowly declining—until around 1900. Then, at the beginning of the 20th century, temperatures began climbing suddenly and swiftly and continued to rise throughout the 1990s, nicely correlating with the growth of modern, industrial, hydrocarbon-burning civilization. As the Intergovernmental Panel on Climate Change (IPCC) famously concluded, the temperature increases during the 20th century were "likely to have been the largest of any century during the past 1,000 years"[1] and the "1990s was the warmest decade and 1998 the warmest year"[2] of the millennium. (See figure 1.)

The Famous Hockey Stick
"Exhibit A" for Global Warming

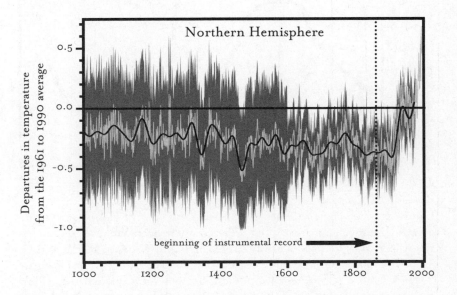

Figure 1. The graph shows not actual temperature for the Northern Hemisphere but how much temperatures in a given period deviated from the 1961 to 1990 average. The smooth black line represents a 50-year moving average; the jagged light gray line the annual average; and the dark gray a 95% confidence range. Note that earlier time periods carry great uncertainty. *Source: Summary for Policymakers* of the 2001 UN IPCC report, p.3. See: http://www.ipcc.ch/pdf/climate-changes-2001/scientific-basis/scientific-spm-en.pdf.

Previous to Mann's work, climate scientists, including those with the IPCC, had a very different idea of temperatures over the past 1,000 years. In 1990, for instance, the IPCC report published the graph showing the more traditional view. That graph showed the Middle Ages to have been quite warm, warmer than today, in what is called the "Medieval Warming Period." Then around the 14th century, we began the descent into the "Little Ice Age," from which we have gradually been emerging since the early 1700s. (See figure 2.)

The Pre-Hockey-Stick View of Global Temperature
Including the Medieval Warming Period

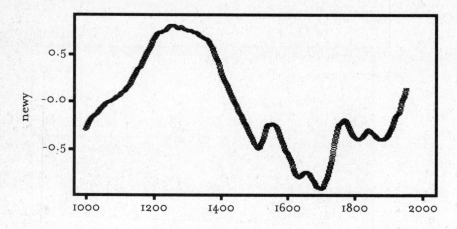

Figure 2. The widely accepted pre-hockey-stick view of global temperature over the last millennium makes late 20th-century temperature change seem less anomalous. *Adapted from* Edward J. Wegman, David W. Scott, and Yasmin H. Said, "Ad Hoc Committee Report on the 'Hockey Stick' Global Climate Reconstruction," p. 34 (presented to the U.S. House of Representatives Committee on Energy and Commerce, July 14, 2006). See: http://www.uoguelph.ca/~rmckitri/research/WegmanReport.pdf.

The hockey stick eliminated the Medieval Warming Period, making our recent 20th-century cooling look far more dramatic compared to the rest of the record. The impact was explosive.

The hockey stick was drawn and redrawn all over the world. More than any other single piece of evidence, it made global warming a serious popular and political issue. And with Mann a lead author of *Climate Change 2001: Third Assessment Report*, the UN IPCC report on global warming, his dramatic graph gained official sanction. Mann's account of what happened to global temperatures over the past thousand years dominated the relevant section of the UN's 2001 *Summary for Policymakers*, always the most influential version of the report, because the authors strip out most of the science so politicians can read the thing.

Precisely because the hockey stick was so dramatic and so often cited, it quickly began to attract critics. The most relentless was a Canadian named Stephen McIntyre. McIntyre was neither a climate scientist nor an eminent scientist of any sort. He was a scientist in the mining industry who recognized the graph's genre the instant he first saw it in 2002—it resembled the deceptive graphics that mining promoters use to hype risky hard-rock mineral exploration projects based on isolated results.

McIntyre may not be an academic scientist, but from the first he proved to be a serious statistician who could ask unsettling questions. He dogged Mann relentlessly, demanding that Mann share more details on his data and methodology. After much back and forth and many struggles to get Mann to identify his data and methods, McIntyre claimed to have proved he found Mann's methodology deeply flawed. In fact, argued McIntyre, when Mann's errors are corrected, the hockey stick disappears. The first graph in figure 3 shows one version of the graph done according to Mann's methodology, with the hockey stick intact. The second shows a reconstruction using the same data, but correcting what McIntyre called Mann's flawed statistics. (See figure 3.)

The Hockey Stick Appears or Disappears Depending on the Statistical Methods Employed

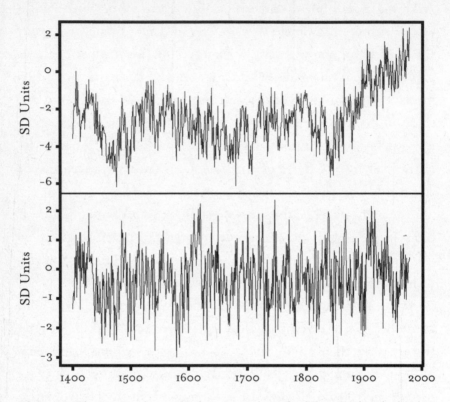

Figure 3. The top graph portrays how Mann's methodology evokes the hockey stick from the tree-ring proxy data. The lower graph shows what happens using the methodology advocated by McIntyre and backed by the Wegman committee and reviewers. *Adapted from* Edward J. Wegman, David W. Scott, and Yasmin H. Said, "Ad Hoc Committee Report on the 'Hockey Stick' Global Climate Reconstruction," p. 32 (presented to the U.S. House of Representatives Committee on Energy and Commerce, July 14, 2006). See: http://www.climateaudit.org/pdf/others/07142006_Wegman_Report.pdf.

See the problem? On McIntyre's showing, not only are the 1990s not necessarily the hottest decade, the 20th century isn't even necessarily the hottest century. The early 1600s are pretty hot, too, and they weren't burning many hydrocarbons back then. That does not prove that burning hydrocarbons is not causing global warming now, any more than Mann's graph proved that current warming was caused by burning hydrocarbons. But the hockey stick's vivid illustration of the coincidence of industrial civilization with 20th-century warming had been an enormously effective prop. Now McIntyre was saying the whole thing was phony, and joining him was Ross McKitrick, an economist at the University of Guelph. In fact, McIntyre and McKitrick claimed Mann's methodology was so flawed that it would produce the same hockey stick shape even with vastly different datasets. To prove it, they ran Mann's methodology with 10,000 simulations using "red noise," or random numbers. Mann's algorithm produced a hockey stick more than 99% of the time.

With the help of the Internet, McIntyre was able to create enough controversy in scientific circles to force Mann to respond. The argument has been going on ever since. Mann has never given in and now has a Web site of his own (www.realclimate.org) largely devoted to attacking McIntyre and McKitrick.

Who is right? McIntyre and McKitrick did ultimately succeed in publishing their critiques of Mann in a peer-reviewed scientific journal.[3] Although Mann was a newly minted Ph.D. when he first drew the hockey stick, he is a specialist in the field of paleoclimatology. McIntyre isn't. The argument from authority seems clear. We have to believe Mann. The UN did, and his standing in the profession demands it.

Enter Dr. Edward Wegman. The hockey stick argument seemed so crucial that the U.S. Congress got involved. Wegman accepted the Energy and Commerce Committee's assignment and agreed to assess the hockey stick controversy pro bono, assembling an expert panel of statisticians to help with the job, also working pro bono.[4] Wegman consulted outside statisticians as well, including the Board of the American Statistical Association. In 2006, the Wegman panel issued its report. The conclusion? The Wegman review repudiated Mann's hockey stick and vindicated his Canadian critics.

"Our committee believes that the assessments that the decade of the 1990s was the hottest decade in a millennium and that 1998 was the hottest year in a millennium cannot be supported,"[5] Wegman's report stated, adding that "the paucity of data in the more remote past makes the hottest-in-a-millennium claims essentially unverifiable."[6] When Wegman corrected Mann's statistical mistakes, the hockey stick disappeared just as McIntyre claimed.

But wait a minute. Mann is a specialist in paleoclimate studies. Wegman is a statistician, admittedly one of the most eminent statisticians in the world, but not primarily a climate scientist. Should we take his word or Mann's?

Statistics are immensely important in climate studies. Because of the paucity of data prior to the mid-19th century, studies of ancient climate are very largely exercises in using advanced statistical techniques to wring useful conclusions out of very limited information.

To understand this, start with an obvious example from modern times. Was the average temperature in the Northern Hemisphere yesterday higher or lower than the day before? Even answering that obvious question requires some sophisticated statistics. Obviously to get our result, we are not going to measure temperature in just one place. We want an average for the entire hemisphere. On the other hand, we are not going to place a thermometer on every square inch of the Northern Hemisphere either. So if one thermometer for the whole hemisphere isn't enough, and one per square inch isn't in the cards, do we need ten thermometers or 10,000 to get a reliable result? How do we figure out how many are enough and where we should place them? Those are largely questions for statisticians. Statistics tells us how much data of what sort we need to answer a given question. Or to flip it around, it can tell us, given a particular set of data, how we can get the clearest picture out of it and how confident we can be that our picture is accurate.

In comparing current temperatures to temperatures a thousand years ago, statistical methods become even more important—as they always do when the data get scarce and patchy. Until relatively recently, we did not have thousands of thermometers distributed around the world, all continuously measuring the same thing. For earlier periods, we depend on sporadic and discontinuous data, such as tree rings, ice cores, lake and ocean sediment, and more. None of these "proxies," as they are called, measure temperature directly. Multiple factors, for instance, influence tree-ring growth. Moreover, these proxy measurements may not be continuously available over the time period. We may have usable tree-ring samples at a given location for half the period in question and none for the rest.

Statistics fill in the gap. This is done, very broadly speaking, by extrapolating from how various proxy measurements correlate with temperature during the period from which we have accurate records to an estimate of what temperatures must have been at times for which we have only proxy data.

Unsurprisingly, this can be a ferociously complicated task. Two top-notch statisticians might well disagree about the best methods to apply in a given case. The decisions on method are rarely straightforward, because different datasets have different shortcomings, and how to compensate for those shortcomings can be a judgment call.

What the Wegman committee found was that the Mann's hockey stick was, just as McIntyre claimed, the result of an error in Mann's methodology. "We explicitly looked at the first principal component of the North American Tree-Ring series and demonstrated that the hockey stick shows up when the data are decentered, but not when properly centered. We also demonstrated the same effect with the digitized version of the 1990 IPCC curve."[7]

Mann's methodology does not create the hockey stick out of whole cloth. Rather, it "mines" the data to find it. If there is a hockey stick in the data somewhere, Mann's algorithm will bring it back alive and magnify its importance. "[T]he decentering process as used in [Mann's papers] selectively prefers to emphasize the hockey stick shape" [and] "will 'data mine' for those shapes.... Most proxies do not contain the hockey stick signal. The [Mann] methodology puts undue emphasis on those proxies that do exhibit the hockey stick shape and this is the fundamental flaw. Indeed, it is not clear that the hockey stick shape is even a temperature signal because all the confounding variables have not been removed."[8]

Wegman argued not only that Mann was wrong but also that the mistakes he made were ones that would have been fairly obvious to a top-notch statistician. Mann and his team had made a basic error that "may be easily overlooked by someone not trained in statistical methodology."[9] Of course Mann and his colleagues all do statistics. To do serious science today often requires a fairly impressive practical mastery of statistics. Wegman's point was that specialists in statistical science "are constantly inventing new methods appropriate to new datasets."[10] Unless paleoclimate experts like Mann brought such statistical specialists into the picture, they risked using misleading methods to analyze challenging data.

Throwing his net a bit wider, Wegman found that this tendency to be behind the curve on statistical issues was endemic in paleoclimate research. The Wegman committee examined lists of references in paleoclimate papers for evidence that the authors were using contemporary statistical tools or citing the current statistics literature. He looked at the résumés of the most frequently published authors in the field to understand where and with whom they obtained their statistical training. He also examined issues such as how many members of the American Meteorological Society's Committee on Probability and Statistics were also members of the prestigious American Statistical Association. The answer at the time: two out of nine.[11]

He concluded that "the atmospheric science community, while heavily using statistical methods, is remarkably disconnected from the mainstream community of statisticians in a way, for example, that is not true of the medical and pharmaceutical communities."[12] Specifically, Wegman pointed out "there is no evidence that Dr. Mann or any of the other authors in paleoclimate studies have had significant interactions with mainstream statisticians."[13]

In short, a renowned authority like Mann, who has done impressive science in his own field, could make an obvious and catastrophic statistical error for the simple reason that he was not in contact with the relevant authorities in statistical science. The argument from authority only works if you actually use the authorities.

All this was devastating enough. But then Wegman asked another question: Why didn't the peer review process work? Why didn't the third-party reviewers catch the errors in Mann's work?

The answer is disturbing, especially considering the great emphasis that the doomsayer faction puts on the existence of scientific consensus. What Wegman discovered is that the real problem in the field is *too much consensus*, especially in the peer-review process. The reviewers all came from the same rather tight circles of the paleoclimate community in which Mann was such a respected figure. Even if personal relationships did not prejudice the outcome of the peer review, the simple discouraging fact was that the reviewers were no more likely to catch Mann's mistake than Mann was, because their own knowledge base was limited pretty much the same as Mann's.

Being a statistician, Wegman did not just assert that paleoclimatologists were too clubby to critique each other; he did the numbers. He and his team did something called "a social network analysis"[14] of the paleoclimate community. Inside that community, according to Wegman's analysis, are several intensively coupled groups. As Wegman explains, statisticians actually call such groups with such tight "complete connections" *cliques*,

THE CV OF A DENIER
Dr. Edward Wegman

Edward Wegman received his Ph.D. degree in mathematical statistics from the University of Iowa. In 1978, he went to the Office of Naval Research, where he headed the Mathematical Sciences Division with responsibility navy-wide for basic research programs. He coined the phrase *computational statistics* and developed a high-profile research area around this concept, which focused on techniques and methodologies that could not be achieved without the capabilities of modern computing resources and led to a revolution in contemporary statistical graphics. Dr. Wegman was the original program director of the basic research program in Ultra High Speed Computing at the Strategic Defense Initiative's Innovative Science and Technology Office. He has served as editor or associate editor of numerous prestigious journals and has published more than 160 papers and eight books.

Repeatedly honored by his peers, Wegman is a member of the Board of The American Statistical Association, a past president of the International Association of Statistical Computing, and a past chairman of the Committee on Applied and Theoretical Statistics of the National Academy of Sciences.

in this case meaning that every member of the group "has one or more coauthored relationships with every other member of the group. . . ."[15]

"In Dr. Mann's case," Wegman continues, "there is exhibited a strong tendency to work with different cliques of closely connected coauthors. . . . It is precisely in a small, specialized discipline that the likelihood of turning up sympathetic referees is highest,"[16] leading to the suspicion that the peer-review process does not fully vet papers before they are published. Noting that "there are a host of fundamental statistical questions that beg answers in understanding climate dynamics,"[17] the hockey stick experience has convinced Wegman that much of climate science should be taken with a grain of salt, since so many studies have been peer reviewed by reviewers unqualified in statistics. Past studies, he believes, should be reassessed by competent statisticians, and in the future the climate-science world should do better at incorporating statistical know-how.

So where does the hockey stick stand today?

The IPCC has dropped it from the *Summary for Policymakers* for its 2007 Report, which is fairly damning. But the hockey stick did its main work years ago and is still very widely cited by advocates of the science-is-settled position.

What does this say about the science being settled? If the hockey stick was at one time so important to the science-is-settled position that the UN was shouting it out from the rooftops, but now it turns out to be phony, does that mean the science is not settled?

CHAPTER THREE

Front Page News

Richard Tol, Christopher Landsea, Duncan Wingham

Wegman's story surprised me. When I started looking for deniers, I expected—or hoped—to find credible scientists who weren't on the take and who could cast some doubt on the forecast of climate-change calamity. Wegman wasn't merely credible. He was at the pinnacle of his academic discipline, the embodiment of the scientific establishment. More surprising, he didn't merely poke holes in some scientific side issue propounded by one of the lesser-known scientists among the 2,500 involved with the UN's Intergovernmental Panel on Climate Change. He took on Michael Mann, perhaps the most prominent and most cited of the IPCC's scientists, and demonstrated before a congressional committee that Mann was utterly out of his depth. Even Mann's illustrious defender in Congress—Gerald R. North of Texas A&M University—confirmed Wegman's critique.

Most surprising of all, Wegman didn't undermine some background study of doubtful significance. He showed the IPCC's single most important climate-change story—the one that garnered the most headlines and won the most public support—to have been entirely without foundation.

Could the other headline cases for global warming—the supposedly rock-solid scientific claims that had made the biggest impression on the public debate—also be wrong? If the public must depend in large part on the argument from authority, it made sense to look into other stories that had made the biggest impression on the public.

In Oct 2006, just as I was getting started on the "Deniers" columns in the *National Post*, huge headlines announced the news that global warming could lead to a global catastrophe "on a scale similar to those associated with the great wars and the economic depression of the first half of the 20th century."[1] The headlines were prompted by an instantly world-famous document called the *Stern Review on the Economics of Climate Change*, sponsored by the UK government and driven by Sir Nicholas Stern, former chief economist of the World Bank. Like Mann's hockey stick, but much more immediate in its effects, the *Stern Review* had an enormous impact on the global warming debate, prompting, for instance, then British prime minister Tony Blair to urge immediate action to stem global warming.

Stern's involvement immediately raised red flags for me not because he isn't a climatologist or any kind of natural scientist but because of his history with the World Bank. Through the Probe International division of my foundation, I have followed the World Bank for more than 20 years and know it to have been wrong time and time again. This is the agency that financed most of the Third World's failed mega-development schemes, that created and perpetuated its debt crises, that engineered the takedown of much of the world's rainforests, and that underwrote the uneconomic dam-building programs that flooded some millions of people off their land, most without compensation.

The World Bank justified the policies that led to all this mayhem on sweeping economic grounds, based on projections from computer models that failed. Had Stern resurrected the use of such failed economic models, I wondered, combining them with suspect climate models to arrive at a sweeping new prescription for the globe? The *Stern Review*'s claims to acting ethically, for the good of humanity, also raised my eyebrows. If Stern was making an economic case for taking action on climate change, why cloud the economics with finger pointing? The answers came soon enough, from economists whose economic and environmental standing have few peers. They didn't mince their words.

The first was Dr. Richard S. J. Tol, one of the world's leading environmental economists and an authority on environmental and social justice, to boot. Stern's doomsday prophecies were "preposterous," according to Tol, portraying the world-famous *Stern Review* as an almost comic mishmash of bad math, bad faith, and worst-case scenarios treated as overwhelming probabilities. Tol's outrage stemmed as much from his concern for the environment as for integrity in the discipline of economics: Stern's *chicken little* study, Tol feared, would act to discredit legitimate concerns over climate change and ultimately prevent needed reforms.

Tol, who holds multiple academic appointments, including those at such prestigious institutions as Hamburg and Carnegie Mellon universities, is no fringe outsider to the scientific debate. Hardly a *denier* in the ordinary sense of the term, he is a central figure in what might be called the "global warming establishment." He is an author with all three Working Groups of the United Nation's Intergovernmental Panel on Climate Change. He is also an author and editor of the United Nations *Handbook on Methods for Climate Change Impact Assessment and Adaptation Strategies*. And he is a mover and shaker in the prestigious European Climate Forum. He takes global warming seriously and has dedicated his professional life to making a contribution for the better in climate policy and related fields.

In addition, he is an expert on the credibility of the *Stern Review* for the excellent reason that some of Stern's most terrifying conclusions were based on Tol's work—which Tol says Stern twisted out of shape to reach absurd conclusions.

Exhibit A: one of Tol's important studies estimated the costs to the global economy of additional CO_2 emissions. Because of the great uncertainty involved, Tol gave a range of possible costs, depending on various assumptions. From this range, Stern plucked a figure of $29 per ton of carbon dioxide without divulging that in the very same study Tol concluded that the actual costs "are likely to be substantially smaller"[2] than $14 per ton of CO_2. Picking the $29 figure, double the $14 figure that Tol said was probably on the high side already, helped the *Stern Review* attain catastrophic conclusions. Likewise, in an assessment of the potential consequences of rising sea levels, Stern quoted a study coauthored by Tol that referred to the "millions at risk,"[3] ignoring that the same study then suggested greatly reduced consequences for those millions due to the proven ability of humans to adapt to change. Strictly speaking, after all, one could say *millions are at risk* this very moment in Tol's native land, the Netherlands, where 60% of the population live below sea level. But the Dutch have coped with this risk for centuries and have attained one of the highest standards of living on Earth, with far less impressive technology than is available today.

Exhibit B: the *Stern Review* claimed that "the overall costs and risks of climate change will be equivalent to losing at least 5% of global GDP each year, now and forever."[4] Not so, says Tol, almost exasperated that claims of this kind can be made, let alone taken seriously. Not only is the "now" an absurd claim, he wrote, "but the 'forever' part is also problematic. It assumes that society will never get used to higher but stable temperatures, changed rainfall patterns, or higher sea levels. This is a rather dim view of human ingenuity. It contradicts what we know about technological progress, adaptation, and evolution."[5]

It is not only Tol's work that Stern distorts by "cherry picking." Tol argues that the *Stern Review*'s assessments of the impact of climate change on water, agriculture, health, insurance, and economic development consistently select the most pessimistic studies in the literature and ignore those "who argue that climate has at most a minor, indirect effect in the (distant) past [or] who show that climate change will have a limited effect on development."[6]

Worse, Tol argues that the economic impact estimates in *Stern* rely on a single (of course, pessimistic) model: "The economic impact estimates of the *Stern Review* are in fact all based on a single integrated assessment model, PAGE2002 by Hope (2006). Although a single model makes for easy presentation, it also implies a lack of robustness."[7] "Integrated assessment models differ considerably in their representation of impacts."[8] "The PAGE2002 model . . . assumes that vulnerability to climate change is independent of development."[9] But we know that rich developed countries deal with natural disasters much better than poor undeveloped countries do. An earthquake in Bangladesh wreaks far more death and human misery than an earthquake of the same intensity in San Francisco.

Tol argues that Stern reaches his fantastic conclusions only by counting up risks not twice, but three times. "The report adds otherwise unspecified and unquantified 'market impacts' (annuitized at 2.1% of GDP), 'non-market impacts' (5.9%), and 'catastrophic risk' (2.9%). . . . The catastrophic risk apart, the *Stern Review* also assumes that there is a 0.1% annual probability (10% in a century) of extinction of the human race. The welfare loss of that is added, too, apparently in addition to the 'catastrophic risk.' Together with the certainty equivalent, risk seems to be counted three times."[10] A 10% per century risk of human extinction does seem a bit high based on historical experience to date.

Tol concludes: "In sum, the *Stern Review* is very selective in the studies it quotes on the impacts of climate change. The selection bias is not random, but emphasizes the most pessimistic studies. . . . Results are occasionally misinterpreted. The report claims that a cost-benefit analysis was done, but none was carried out. The *Stern Review* can therefore be dismissed as alarmist and incompetent."[11]

Tol has eminent company in dismissing *Stern*, including Dr. William Nordhaus, Sterling professor of economics at Yale University, who sits on the National Academy of Sciences Committee

THE CV OF A DENIER
Dr. Richard Tol

Richard Tol received his Ph.D. in economics from the Vrije University in Amsterdam. He is the Michael Otto Professor of Sustainability and Global Change and director of the Center for Marine and Atmospheric Sciences at Hamburg University, principal researcher at the Institute for Environmental Studies at Vrije University, and adjunct professor at the Center for Integrated Study of the Human Dimensions of Global Change at Carnegie Mellon University. He is a board member of Hamburg University's Center for Marine and Climate Research, the International Max Planck Research Schools on Earth System Modeling and Maritime Affairs, and the European Forum on Integrated Environmental Assessment. He is an editor of *Energy Economics*, an associate editor of *Environmental and Resource Economics*, and a member of the editorial board of *Environmental Science and Policy* and *Integrated Assessment*.

on the Policy Implications of Greenhouse Warming. Yet the eminent dismissals counted for naught in informing public opinion. The worldwide press covered the *Stern Review* lavishly and buried its reviews.

Mann's hockey stick was an essential building block in what the doomsayer camp likes to call the "growing consensus" that global warming is imminent, human-caused, and quite likely catastrophic. The *Stern Review* was powerful in a different way—it undercut political leaders who advocated a cautious, balanced approach to the problem of the sort advocated by Tol. In terms of building popular hysteria, however, the great hurricane headline scare of 2004 and 2005 may have trumped them both.

Stormy Weather

With the 2006 and 2007 Atlantic hurricane seasons having been relatively calm, we hear less often today of the "scientific consensus" that the violent seasons of 2004 and 2005, culminating in Katrina, were caused by global warming. But it was a very popular claim for a while. What dumbfounded me when I profiled my next denier was learning that the hurricane claim was contrived by UN climate-change scientists who had every reason to know it was false—because they were warned in advance by the UN's own leading expert on the subject of Atlantic hurricanes, Dr. Christopher Landsea, of the Atlantic Oceanographic & Meteorological Laboratory.

Landsea was a contributing author for the second UN IPCC report in 1995, writing the sections on observed changes in tropical cyclones around the world. Then the IPCC called on him again as a contributing author for its *Third Assessment Report* in 2001. And he was invited to participate yet again when the IPCC called on him to be an author in the *Fourth Assessment Report*, to be published by the IPCC in 2007, which would specifically focus on Atlantic hurricanes, Landsea's specialty.

Then something very odd happened. Within days of this last invitation, in October 2004, Landsea was told that the IPCC's Kevin Trenberth—the very person who had invited Landsea to serve again as an IPCC author—was participating in a press conference at the Harvard Medical School's Center for Health and the Global Environment. The headline on the press release read: "EXPERTS TO WARN GLOBAL WARMING LIKELY TO CONTINUE SPURRING MORE OUTBREAKS OF INTENSE HURRICANE ACTIVITY."[12] Landsea was shocked. Not only did Landsea's work not substantiate the claim that global warming was causing more or worse hurricanes, as far as he knew, no research substantiated such a claim. All previous and current research in the area of hurricane variability, Landsea recognized, showed no reliable upward trend in the frequency or intensity of hurricanes. Not in the Atlantic basin. Not in any other basin.

Moreover, the IPCC itself, in both 1995 and 2001, had found no global warming signal in the hurricane record. And until the new report, which Landsea had been asked to write, was completed in 2007, the IPCC would not have a new analysis on which to base this new claim that hurricanes were getting worse.

Landsea e-mailed Trenberth and another colleague, Dr. Linda Mearns, who had originally been announced as a participant in the press conference, to express his concerns that the press conference could stray from science into hyperbole. Noting the lack of research to support the headline claim in the press release, Landsea wrote, "I am wondering what has led you all to this big conclusion. Are you all announcing a new published paper in the field?" noting that he had not seen any work on the subject from Trenberth, Mearns, or any other member of the panel. . . . "If not," Landsea continued, "I'm wondering how you all are coming to this conclusion and what work you are citing?"[13] He then offered his own synopsis of the state of knowledge on hurricanes and global warming (reproduced in full on pages 38–39), offering to provide full references if his colleagues were interested. Summed up in a single sentence: "There are no known scientific studies that show a conclusive physical link between global warming and observed hurricane frequency and intensity."[14]

Dr. Mearns withdrew from the press conference, though Landsea does not know whether she did so in reaction to his e-mail. Otherwise the event went on as planned. The press release, including statements from Dr. Trenberth and the other scientists, was a masterpiece of equivocation. The wording might almost have been deliberately contrived to cause maximum hysteria in the media, while not actually committing any of the scientists to the blunt claim that *global warming causes hurricanes or makes them worse*.

"With four hurricanes and tropical storms hitting the United States in a recent five-week period, 2004 already is being called 'The Year of the Hurricane,'" the press release led off. "But this year's unusually intense period of destructive weather activity could be a harbinger of what is to come as the effects of global warming become even more pronounced in future years, according to leading experts who participated today in a Center for Health and the Global Environment at a Harvard Medical School briefing."[15] This paragraph never commits anyone to any definite hypothesis. "Could be" covers a multiple of possibilities. But it gave the press something to run with. Perhaps even more masterful was Trenberth's statement:

"Human activities are changing the composition of the atmosphere and global warming is happening as a result," says Kevin Trenberth, head of the Climate Analysis Section at NCAR and a convening lead author of the 2007 IPCC report for the chapter on observed changes. "Global warming is manifested in many ways, some unexpected. . . . The environment in which hurricanes form is changing. The result was a hurricane in late March 2004 in the South Atlantic, off the coast of Brazil: the first and only such hurricane in that region. Several factors go into forming hurricanes and where they track. But the evidence strongly suggests more intense storms and risk of greater flooding events, so that the North Atlantic hurricane season of 2004 may well be a harbinger of the future."[16]

He never quite said it, but the message was clear: *global warming is a global terror and we are already seeing its catastrophic consequences.*

Hurricanes had been all over the news that summer. And some in the media were already speculating about whether global warming was the culprit. Trenberth's press conference was perfect for that story, with the climate-change experts at hand all apparently (but not quite) confirming the news that the public had been primed to hear: *global warming was causing hurricanes.*

The press conference made headlines all around the world. The coverage from Reuters was typical:

Global Warming Effects Faster Than Feared – Experts
Thu Oct 21, 2004 3:32 PM ET Science
By Maggie Fox, Health and Science Correspondent
    WASHINGTON (Reuters)—Recent storms, droughts, and heat waves are probably being caused by global warming, which means the effects of climate change are coming faster than anyone had feared, climate experts said on Thursday.
    The four hurricanes that bashed Florida and the Caribbean within a five-week period over the summer, intense storms over the western Pacific, heat waves that killed tens of thousands of Europeans last year, and a continued

drought across the U.S. southwest are only the beginning, the experts said . . .[17]

Amazed, Landsea wrote to top IPCC officials, protesting:

"Where is the science, the refereed publications, that substantiates these pronouncements? What studies are being alluded to that have shown a connection between observed warming trends on the Earth and long-term trends in tropical cyclone activity? As far as I know, there are none. . . . There are many legitimate scientific reasons to be concerned with global warming, but the evidence just is not there with hurricanes no matter how much it is trumped up for the media and the public. Proceeding with such announcements outside the proper IPCC process taints the credibility of climate-change science and will in the longer term diminish our influence over public policy."[18]

Landsea asked the IPCC leadership for assurances that the IPCC's 2007 report would be true to science: "The lead author [Dr. Trenberth] seems to have already come to the conclusion that global warming has altered hurricane activity and has publicly stated so. This does not reflect the consensus within the hurricane research community. . . . Thus I would like assurance that what will be included in the IPCC report will reflect the best available information and the consensus within the scientific community most expert on the specific topic."[19]

The assurance did not come. The relevant IPCC leadership, to the extent they responded at all, took the position that since the Harvard press conference was not an official IPCC event, Trenberth had the right to say whatever he pleased and the IPCC had no business commenting, never mind disciplining or even remonstrating with Trenberth, in any way. Dr. R. K. Pachauri, the IPCC's chairman, went even further, arguing that Trenberth "did not in any way misrepresent the IPCC and apparently his statements accurately reflected IPCC's *TAR* [*Third Assessment Report*, issued in 2001]."[20]

Landsea shot back, saying that whether the IPCC officially backed the Harvard press conference was beside the point. Here we had a lead IPCC author publicly declaring his views on research not yet completed, his views flying in the face of the known science, and publicizing these views in a way apparently calculated to spread hysteria and implant in the public ideas known to be contrary to the scientific evidence.

Moreover, argued Landsea, Trenberth's statements certainly were not consistent with the *TAR*, which Landsea helped write. The *TAR Summary for Policymakers* had been extremely specific: "Some important aspects of climate appear not to have changed . . . Changes globally in tropical and extra-tropical storm intensity are dominated by inter-decadal to multi-decadal variations, with no significant trends over the 20th century."[21] In the more technical section of the *TAR*, the most the IPCC scientists had been willing to say, Landsea pointed out, was that it was "possible that tropical cyclones will develop slightly stronger winds and rain in the somewhat distant future,"[22] i.e., about 80 years from now according to recent research. This, he pointed out, is far from saying that 2004's catastrophic storms were caused by global warming.

Convinced that the IPCC in backing Trenberth was corrupting science, Landsea resigned. His resignation letter makes clear that it was the IPCC's failure to see Trenberth's actions as problematic that convinced him the IPCC itself was compromised:

> The IPCC leadership saw nothing to be concerned with in Dr. Trenberth's unfounded pronouncements to the media, despite his supposedly impartial important role. . . .
> Because of Dr. Trenberth's pronouncements, the IPCC

process on our assessment of these crucial extreme events in our climate system has been subverted and compromised, its neutrality lost. While no one can 'tell' scientists what to say or not say (nor am I suggesting that), the IPCC did select Dr. Trenberth as a Lead Author and entrusted to him to carry out this duty in a non-biased, neutral point of view. . . . It is of more than passing interest to note that Dr. Trenberth, while eager to share his views on global warming and hurricanes with the media, declined to do so at the Climate Variability and Change Conference in January where he made several presentations. Perhaps he was concerned that such speculation—though worthy in his mind of public pronouncements—would not stand up to the scrutiny of fellow climate scientists.

I personally cannot in good faith continue to contribute to a process that I view as both being motivated by preconceived agendas and being scientifically unsound. As the IPCC leadership has seen no wrong in Dr. Trenberth's actions and have retained him as a Lead Author for the *AR4*, I have decided to no longer participate in the IPCC *AR4*.[23]

In March 2007, almost two and a half years after Landsea's unsuccessful attempt to stop the press conference that tied global warming to hurricanes, the IPCC released its *Fourth Assessment Report*. The IPCC now downplayed hurricanes—the 2006 hurricane season failed to live up to its billings. The *Summary for Policymakers* stated: "There is no clear trend in the annual numbers of tropical cyclones."[24] Almost without exception, the media outlets that had once so voraciously covered the IPCC's hurricane warnings fell still. Hardly anyone noticed that the dire warnings on hurricanes failed to materialize, as might well have been expected had Landsea been

THE CV OF A DENIER
Dr. Christopher Landsea

C hristopher Landsea received his doctoral degree in atmospheric science from Colorado State University. A research meteorologist at the Atlantic Oceanographic and Meteorological Laboratory of the National Oceanic and Atmospheric Administration, he was chair of the American Meteorological Society's Committee on Tropical Meteorology and Tropical Cyclones and a recipient of the American Meteorological Society's Banner I. Miller Award, recognized for the "best contribution to the science of hurricane and tropical weather forecasting." He is a frequent contributor to leading journals including *Science, Bulletin of the American Meteorological Society, Journal of Climate,* and *Nature.*

listened to. Like the Mann hockey stick, the IPCC quietly banished hurricanes as cover-story material. Also like the Mann hockey stick, the hurricane fears have done their work. Even if they are no longer top of mind, they remain in the public's consciousness, to be called up whenever a new hurricane or even hurricane warning is announced on a TV or radio newscast.

Polar scientists on thin ice

A great melt is on in Antarctica. Its northern peninsula—a jut of land extending to about 1,200 kilometers from Chile—has seen a drastic increase in temperature, a thinning of ice sheets, and most alarmingly, a collapse of ice shelves. The Larsen A ice shelf, 1,600 square kilometers in size, fell off in 1995. The Wilkins ice shelf, 1,100 square kilometers, fell off in 1998, and the Larsen B,

13,500 square kilometers, dropped off in 2002. (For comparison, the state of Connecticut covers about 14,400 square kilometers.) Meanwhile, the northern Antarctic Peninsula's temperatures have soared by 6°C in the last 50 years.

This melting has been a huge part of the doomsayer case for global warming. Al Gore depends heavily on it. For years, we have seen headlines on the order of "Antarctic Being Destroyed," "Antarctic Ice Sheet Melting Rapidly," and of course, "Penguins Threatened by Global Warming."

There is good reason for this. As Richard Tol points out, Antarctica represents the greatest threat to the globe from global warming, bar none. If Antarctica's ice melts, the world's oceans will rise, flooding low-lying lands where much of the world's population lives, including many millions of people from under-developed countries in the Southern Hemisphere least able to cope with the crisis. Not only could millions of people be displaced from their homes, the world would lose fertile deltas that feed tens of millions of people. This chilling scenario understandably sends shudders through concerned citizens around the world and steels the resolve of those determined to stop the cataclysm.

Fortunately, according to some of the world's leading experts on the question, the melting of Antarctica *may* be as much of a myth as the great hurricane scare. Oh, there is no doubt we are losing ice on the northern peninsula, a tiny sliver of the continent stretching northward beyond the Antarctic Circle. But Antarctica is a big place, covering about 14-million square kilometers altogether, most of it much less accessible and much less studied than the northern tip. Taking the entire continent into consideration, much confounding evidence exists. As one example, at the South Pole, where the United States decades ago established a polar research station, temperatures have actually fallen since 1957.

Enter Duncan Wingham, professor of Climate Physics at University College London and director of the Center for Polar Observation and Modelling. Dr. Wingham has been pursuing this polar puzzle for much of his professional life and, but for an accident in space, he might have had the answer at hand by now.

Dr. Wingham is principal scientist of the European Space Agency's CryoSat Mission, a $130-million project designed to map changes in the depth of ice using ultra-precise instrumentation. Sadly, for Dr. Wingham and for science as a whole, CryoSat fell into the Arctic Ocean after its launch in October 2005, when

Hurricanes and Global Warming

From Dr. Christopher Landsea's October 21, 2004,
e-mail message to Kevin Trenberth of the IPCC et al.

- There are no known scientific studies that show a conclusive physical link between global warming and observed hurricane frequency and intensity. Whatever suggested changes in hurricane activity that might result from global warming in the future are quite small in comparison to the large, natural variability of hurricanes, typhoons, and tropical cyclones. For example, the latest GFDL [Geophysical Fluid Dynamics Laboritory] global warming study suggested about a 5% increase in the winds of hurricanes 80 years in the future. This contrasts with the more than doubling that occur[s] now in numbers of major hurricanes between active and quiet decades in the Atlantic basin.

- [If] global warming is influencing hurricane activity, then we should be seeing a global change in the number and strength of these storms. Yet there is no evidence of a global increase in the strength and frequency of hurricanes, typhoons, and tropical cyclones over the past several years.

- Beginning in 1995, there has been an increase in the frequency a rocket launcher malfunctioned. Dr Wingham will now need to wait until 2009 before CryoSat II, CryoSat's even more precise successor, can launch and begin relaying the data that should conclusively determine whether Antarctica's ice sheets are thinning or not. Apart from satellite technology, no known way exists to reliably determine changes in ice mass over a vast and essentially inaccessible continent covered in ice several kilometers thick.

Fortunately, CryoSat was not the only satellite available to polar scientists. Dr. Wingham has been collecting satellite data for years and arriving at startling conclusions. In February 2005, at a

and intensity of hurricanes in the Atlantic basin. However, this increase is very likely a manifestation of a natural multi-decadal cycle of Atlantic hurricane activity that has been occurring likely for the past few hundred years. For example, relatively few Atlantic major hurricanes were observed in the '70s, '80s, and early '90s, but there was considerable activity during the '40s, '50s, and early '60s. Also, the period from 1944 to 1950 was particularly infamous for Florida—with 11 hurricanes hitting the state during those years.

- Total U.S. direct damages from Atlantic hurricanes this year [2004] will be on the order of $30 billion, making it about equal to the most damaging year on record—1992, with the landfall of Hurricane Andrew. However, such increased destruction from hurricanes is to be expected because of the massive development and population increases along the U.S. coastline and in countries throughout the Caribbean and Central America. There is no need to invoke global warming to understand both the ten years of active hurricane seasons and the destruction that occurred both in Florida and in Haiti this season. The former is due to natural cycles driven by the Atlantic Ocean and the latter is due to societal changes, not due to global warming.

European Union Space Conference in Brussels, for example, Dr. Wingham revealed that data from a European Space Agency satellite showed that Antarctic thinning was no more common than thickening. But what about the spectacular collapse of the ice shelves? In Wingham's view, this was much more likely to have been driven by natural current fluctuations than by global warming. In short, the great ice shelf collapse *may have been* much like the great hurricane scare. A catastrophic event occurs coincident with an intense debate over global warming and then gets attributed to that warming not so much because of hard evidence but because it makes a great story.

As Wingham explained in an extensive interview on the topic, "A lot of attention and research has focused on this relatively accessible area of the Antarctic Peninsula," in part because the climate is moderate. This emphasis is misleading. "The Antarctic Peninsula is exceptional because it juts out so far north."[25] Meanwhile, "satellites are giving us a picture of the continent as a whole."[26] Because many scientists have been preoccupied with what was, in effect, the tip of the iceberg, they missed the mass of evidence that lay beneath the surface.

Wingham, far from a strident denier, is cautious. We cannot be certain, he explains, that the collapse of the northern ice shelves is not the result of human-caused global warming, "because packets of heat in the atmosphere do not come conveniently labeled 'the contribution of anthropogenic warming.'" But, he argued, the evidence is not "favorable to the notion we are seeing the results of global warming."[27]

He stresses that he is "not denying global warming. For instance, Greenland, in the Northern Hemisphere, does seem to be going. But Greenland's ice cap—Greenland is quite far south—is a last survivor from the Ice Age and only its height protects it. The more that cap melts, the more it will continue to melt as it

THE CV OF A DENIER
Dr. Duncan Wingham

Duncan Wingham was educated at Leeds and Bath universities where he gained a B.Sc. and Ph.D. in physics. He was appointed to a chair in the Department of Space and Climate Physics, University College London, in 1996, and to head of the Department of Earth Sciences in October 2005. Prof. Wingham is a member of the National Environmental Research Council's Science and Technology Board and Earth Observation Experts Group. He is a director of the NERC Center for Polar Observation & Modelling and principal scientist of the European Space Agency CryoSat Satellite Mission, the first ESA Earth Sciences satellite selected through open scientific competition.

gets lower and warmer." He goes so far as to say that "no one doubts" that Arctic ice faces a grim fate, though he cautions that actual data are sparse and that satellite data show both growth and melting up north. But, he says, "Antarctica is different."[28]

In 2006, Dr. Wingham and three colleagues published an article in the journal of the Royal Society that casts further doubt on the notion that global warming is adversely affecting Antarctica. By studying satellite data from 1992 to 2003 that surveyed 85% of the East Antarctic ice sheet and 51% of the West Antarctic ice sheet (72% of the ice sheet covering the entire land mass), they concluded that the Antarctic ice sheet is growing at the rate of 5 millimeters per year (plus or minus 1 mm per year). That makes Antarctica a sink, not a source, of ocean water. According to their best estimates at the time, Antarctica will "*lower* [authors' italics] global sea levels by 0.08 mm"[29] per year.

Five millimeters per year of growth translating into eight one-hundredths of a millimeter per year of dropping sea level is a trivial amount. The following year, based on additional research, it appeared to Wingham and colleagues that the ice sheet might be shrinking—again by a trivial amount.

CryoSat II and other research developments should allow scientists to assess the 28% of Antarctica not yet surveyed, giving a more definitive picture of what's happening. If those results confirm Dr. Wingham's findings to date—that the massive Antarctic continent is also, on balance, massively stable—the low-lying areas of the world will have nothing to fear from Antarctica. This finding would certainly not surprise him—it would merely be a continuation of a century-long freeze in the status quo. "At most, Antarctica has had only a modest (+/- 1–2 mm) impact upon global sea levels this century,"[30] Dr. Wingham observes.

CHAPTER FOUR

Forging A Consensus

Robert Carter, Richard Lindzen

I have to admit to being shocked by the stories of the first four deniers I wrote about—Wegman, Tol, Landsea, and Wingham. I had picked several of the most essential and/or most widely publicized "building blocks" of the case for catastrophic global warming. In each case, not only was I able to find a truly eminent, world-renowned leader in the field who disputed the point in question, but in each case the denier had more authority, sometimes far more authority, than those who put forward the building block in the first place.

This is true straight on down the line, starting with the first and hardest case. Mann clearly commands enormous respect among his colleagues in paleoclimatology. I do not quarrel with his reputation as a scientist. His most visible and persistent critics, McIntyre and McKitrick, do not have anything like Mann's credentials in climatology.

The fact remains that, because the data on pre-19th century temperature is so sparse, the reconstruction of early temperatures is largely an exercise in statistical science. And on statistical issues, not only Wegman but the panel of experts he assembled carries incomparably more weight than Mann. And the authorities in statistics rule out the methods Mann used to evoke the hockey stick. With the hockey stick dead as evidence, dead, too, is the single, indispensable, underlying argument that recent warming has been so unusual over the last millennium as to demand an explanation beyond natural forces. We do not need to rewrite the history books, as would have been required had Mann been correct. Science does not tell us that we live in warmer times now than we did prior to the Little Ice Age, roughly the centuries immediately preceding Christopher Columbus's discovery of America.

Neither can we unquestioningly accept Stern's view of the world. Stern based his conclusions, in good part, on Tol and on a field of research that Tol knows far better than Stern. Moreover, though I have focused on Tol, Stern has many more critics, among them other eminent experts. If the experts on whom Stern built his case say he got it wrong, what's left of Stern?

The Landsea case is not even close. Landsea is the expert on Atlantic hurricanes, vouched for by the very IPCC officials he criticizes. Moreover, the 2007 IPCC report effectively backs Landsea, just as previous IPCC reports did. This one was a pure case of hype and headlines vs. actual expertise.

Wingham, a careful scientist, is not yet ready to claim for sure whether the Antarctic ice sheet is growing or shrinking. Perhaps Cryosat II will provide definitive evidence of shrinkage. But, even if it is shrinking, he states the cause is likely not global warming but natural. Most of all, his story does show that, as with the hurricane scare, the evidence for the doomsday view is partial, anecdotal, and even media-driven, while the more serious and comprehensive scientific work shows that the science is anything but *settled*.

As these rather dramatic reversals for the doomsday view mounted, however, I also noticed something striking about my growing cast of deniers.

None of them were deniers.

Wegman did not dispute that manmade global warming was occurring. He did not question that the Arctic and Antarctic were melting, that hurricanes were increasing in number and intensity, or that the economic costs of global warming could be crippling. He merely stated that in the one narrow area in which his expertise applied to the public debate over global warming—the use of statistics to support claims of global warming—the claims were unsupportable.

Likewise, Tol is hardly a denier: he was one of the first to identify the economic costs of global warming and is a major force in what might be called the global warming community. Tol does not dispute that hurricane activity has worsened because of global warming, or that sea levels could rise significantly—he has even studied the economic consequences of such occurrences. He merely stated that in the economic arena with which he is so familiar alarmism of the type propounded by Stern could not be justified. Yes, there are economic costs to global warming, he believes, but there are also economic benefits that mitigate some of the costs. Yes, there are net costs, but these appear modest and do not warrant crash programs to cut back on greenhouse gas emissions. To Tol, global warming is as much an issue of social justice as economics, particularly since we in the rich northern nations will often benefit from global warming, while those in the poor southern countries may be losers. Tol's approach: compensate people in the Third World for the damage we do them and open our arms to the millions who want to emigrate to the rich countries.

Landsea, too, has issued no sweeping denials of coming catastrophic global warming. No naysaying to reports that the polar regions are melting due to manmade activities, no questioning of the economic costs of global warming, no doubts that temperatures are higher today than at any time in the last 1,000 years. He merely says that in the one area he knows something about—hurricane activity—there is absolutely no reason to think that global warming is a factor of any significance at all.

Wingham is likewise not a denier. He does not doubt that manmade global warming increases hurricane activity or historic temperatures or economic costs. He merely states that in the one area he knows something about—the polar regions—the doubts are immense. It is not possible to attribute melting here to manmade causes with any degree of confidence.

In effect, all four scientists were saying, "I'm sure global warming exists. All the science from all the different scientific disciplines says so. But there is one exception—my particular area of expertise has found no compelling evidence of manmade global warming."

Affirmers in general. Deniers in particular. Like other smart people, and like most everyone, scientists accept the conventional wisdom in areas they know little about. Put another way, people are predisposed to accept what they believe to be a consensus. We know from our daily lives that the consensus can sometimes be spectacularly wrong. Juries have been known to convict innocent men, as DNA evidence is now making clear. Investors have been known to be irrationally exuberant, as those who got caught by the dot-com bubble learned. A scientific consensus can be no different.

More to the point, the notion of a scientific consensus can mislead. The hallmark of a scientific proposition is that it be testable. And the notion of testability makes notions like consensus quite unnecessary.

Dr. Robert M. Carter is a research professor at University of Adelaide (South Australia) and James Cook University (Queensland), where he headed the School of Earth Sciences between

THE CV OF A DENIER
Dr. Robert Carter

Robert Carter is a research professor at University of Adelaide (South Australia) and James Cook University (Queensland), where he headed the School of Earth Sciences between 1981 and 1999. He has published more than a hundred papers in international science journals and received numerous awards and prizes from bodies such as the Australian Research Council, the Geological Society of New Zealand, and the Royal Society of New Zealand. He received his Ph.D. in paleontology in 1968 from the University of Cambridge.

1981 and 1999. Carter is an eminent scientist in his own right, having published more than 100 papers in scientific journals around the world. But he is a valuable source for another reason as well. He is a clear writer who in addition to his own research has done a masterful survey of the global warming literature. Dr. Carter is particularly eloquent on the subject of scientific consensus. As he points out, "we do not usually say that 'there is a consensus that the sun will rise tomorrow.' Instead, the confident statement that 'the sun will rise tomorrow' rests on repeated empirical testing and the understanding conferred by Copernican and Newtonian theory."[1]

Grand theories, of the sort that usually get named after grand theorists like Copernicus and Newton, may appear as the result of a single, blinding insight dramatically advancing our understanding. But even in their day, to establish the grand theory in the first place and then to extend, test, confirm, and apply it, typically requires the work of specialists working on relatively narrow problems. It is within the narrow confines of their specialties, where they may be truly called experts, that most scientists do real science. The further they stray from what they know best, the more likely they are to be doing things like "forming a consensus," which is not a scientific activity.

As Carter brilliantly points out, this picture of science as necessarily specialized is especially true of climate science, which as yet lacks its "grand theory":

"Much public discussion on global warming is underpinned by two partly self-contradictory assumptions. The first is that there is a 'consensus' of qualified scientists that dangerous human-caused global warming is upon us; and the second is that although there are 'two sides to the debate,' the dangerous-warming side is overwhelmingly the stronger. Both assertions are unsustainable. The first because science is not, nor ever has been, about consensus, but about experimental and observational data and testable hypotheses. Second, regarding the number of sides to the debate, the reality is that small parts of the immensely complex climate system are better or less understood—depending upon the subject—by many different groups of experts. No one scientist, however brilliant, 'understands' climate change, and there is no general theory of climate nor likely to be one in the near future. In effect, there are nearly as many sides to the climate-change debate as there are expert scientists who consider it."[2]

There are nearly as many sides to the climate-change debate as there are expert scientists who consider it. My own very early columns certainly supported the idea that there were very many "sides," not necessarily in the sense of opposed sides, pro and con, but sides in the sense of aspects of the question. And since the aspects of the question are so numerous and complicated, even most scientists specializing in climate or climate-related research cannot be experts on more than a small part of the picture.

Dr. Richard Lindzen has been eloquent on this point. "[I]t is safe to say that global warming consists in so many aspects, that widespread agreement on all of them would be suspect *ab initio*. If it truly existed, it would be evidence of a thoroughly debased field. In truth, neither the full text of the IPCC documents nor even the summaries claim any such agreement."[3]

Like Tol, Dr. Lindzen is a critic from within. He is one of the most distinguished climate scientists in the world: a past professor at the University of Chicago and Harvard; the Alfred P. Sloan professor of meteorology at the Massachusetts Institute of Technology; a member of the National Academy of Sciences; and a lead author in the landmark 2001 UN IPCC report. Lindzen readily acknowledges that temperatures have risen in the 20th century along with CO_2 levels and that increased CO_2 can raise temperature. But none of those well-established facts, he argues, can justify the doomsday case, or the doomsday solution, and certainly cannot justify twisting science for political goals.

Lindzen remains proud of his contribution, and that of his colleagues, to the IPCC chapter 7 they worked on entitled "Physical Processes." His pride in this work matches his dismay at seeing it misrepresented. "[A]lmost all reading and coverage of the IPCC is restricted to the highly publicized *Summaries for Policymakers*, which are written by representatives from governments, NGOs, and business; the full reports, written by participating scientists, are largely ignored,"[4] he told the United States Senate Committee on Environment and Public Works in 2001. These unscientific summaries, possibly driven by political or business agendas, then become the basis of public understanding.

As an example, Dr. Lindzen offers the summary that was created for the chapter of the 2001 report he helped author. "Understanding of climate processes and their incorporation in climate

THE CV OF A DENIER
Dr. Richard Lindzen

Richard Lindzen received his Ph.D. in applied mathematics in 1964 from Harvard University. A past professor at the University of Chicago and Harvard, he is the Alfred P. Sloan professor of meteorology at the Massachusetts Institute of Technology. He is a member of the National Academy of Sciences, a fellow of the American Association for the Advancement of Science, and a member of the National Research Council Board on Atmospheric Sciences and Climate. He is also a consultant to the Global Modeling and Simulation Group at NASA's Goddard Space Flight Center and a Distinguished Visiting Scientist at California Institute of Technology's Jet Propulsion Laboratory. Professor Lindzen is a recipient of the American Meteorological Society's Meisinger and Charney Awards and American Geophysical Union's Macelwane Medal. He is author or coauthor of over 200 scholarly papers and books.

models have improved, including water vapor, sea-ice dynamics, and ocean-heat transport,"[5] the summary stated, creating the impression that the climate models were reliable. The actual report by the scientists indicated just the opposite. Dr. Lindzen testified that the scientists had "found numerous problems with model treatments—including those of clouds and water vapor."[6] As Lindzen argued in a now famous April 12, 2006, *Wall Street Journal* article titled "Climate of Fear": "[I]t isn't just that the alarmists are trumpeting model results that we know must be wrong. It is that they are trumpeting catastrophes that couldn't happen even if the models were right."[7]

The IPCC was stung by criticism that the summaries were being written with little or no input by the scientists themselves. So the IPCC created a new stage in the review process in which selected scientists reviewed a subsequent draft summary. Unfortunately, when the final version was later released at a Shanghai press conference, it had surprising changes to the draft that scientists had originally seen.

The version that emerged from Shanghai concluded, "In the light of new evidence and taking into account the remaining uncertainties, most of the observed warming over the last 50 years is likely to have been due to the increase in greenhouse gas concentrations."[8] This is an extremely powerful claim. It suggests that CO_2 and other greenhouse gases have a much more powerful effect on climate than had been generally believed and implies quite large and rapid increases in temperature unless we drastically reduce emissions. But this dramatic claim turns out to be drama by distortion. The draft reviewed by the scientists had been rife with qualifiers, making it clear the science was very much in doubt because "the accuracy of these estimates continues to be limited by uncertainties in estimates of internal variability, natural and anthropogenic forcing, and the climate response to external forcing."[9]

Dr. Lindzen's description of the conditions under which the climate scientists worked conjures up a scene worthy of a totalitarian state: "Throughout the drafting sessions, IPCC 'coordinators' would go around insisting that criticism of models be toned down, and that 'motherhood' statements be inserted to the effect that models might still be correct despite the cited faults. Refusals were occasionally met with *ad hominem* attacks. I personally witnessed coauthors forced to assert their 'green' credentials in defense of their statements."[10]

In his Senate testimony, Lindzen noted that both the media

and the politicians often misunderstood and misused the summaries. But he argued that the IPCC is just as much at fault for following practices that "encourage misuse" and:

- Use a summary to misrepresent what scientists say.
- Use language which conveys different meaning to laymen and scientists.
- Exploit public ignorance (and the embarrassment about this ignorance) over quantitative matters.
- Exploit what scientists can agree on in order to support one's agenda.
- Exaggerate scientific accuracy and certainty.
- Exaggerate the authority of undistinguished scientists.[11]

"So how is it that we don't have more scientists speaking up about this junk science?"[12] he asks. His grim answer: carrots and sticks. Those who toe the party line are publicly praised and have grants ladled out to them from a funding pot that overflows with more than $1.7 billion per year in the United States alone. Those who don't are subject to attack.

As far back as 1992, Lindzen explained, then-senator Al Gore "ran two congressional hearings during which he tried to bully dissenting scientists, including myself, into changing our views and supporting his climate alarmism."[13] Since then, he says, scientists "who dissent from the alarmism have seen their grant funds disappear, their work derided, and themselves libeled as industry stooges, scientific hacks, or worse. Consequently, lies about climate change gain credence even when they fly in the face of the science that supposedly is their basis."[14] Not hesitating to name names, Lindzen claimed that "Henk Tennekes (see chapter eight) was dismissed as research director of the Royal Dutch Meteorological Society after questioning the scientific underpinnings of global warming. Aksel Winn-Nielsen, former director of the UN's World Meteorological Organization, was tarred by Bert Bolin, first head of the IPCC, as a tool of the coal industry for questioning climate alarmism. Respected Italian professors Alfonso Sutera and Antonio Speranza disappeared from the debate in 1991, apparently losing climate-research funding for raising questions."[15]

Charges of this nature are almost always difficult, even impossible, to demonstrate. It is a rare grant rejection letter or pink slip that reads *we have no use for you because you are unwilling to toe the*

Richard Lindzen on Stormy Weather

If the models are correct, global warming reduces the temperature differences between the poles and the equator. When you have less difference in temperature, you have less excitation of extra-tropical storms, not more. And, in fact, model runs support this conclusion. Doomsayers have drawn some support for increased claims of tropical storminess from a casual claim by Sir John Houghton of the UN's Intergovernmental Panel on Climate Change (IPCC) that a warmer world would have more evaporation, with latent heat providing more energy for disturbances. The problem with this is that the ability of evaporation to drive tropical storms relies not only on temperature but humidity as well and calls for drier, less humid air. Claims for starkly higher temperatures are based upon there being more humidity, not less—hardly a case for more storminess with global warming.

Richard Lindzen, "Climate of Fear," April 12, 2006

party line. One man's intimidation is another's *frank discussion.* Did the staff rewrite the summaries? Surely only after *a careful review of all contending viewpoints.* But as Bob Carter notes, credible accusations of IPCC bias have also been made by the doomsayer camp:

"Significantly, the recent release of the *Fourth Assessment Report (4AR;* IPCC, 2007) was greeted by strong criticisms also from supporters of the dangerous warming case; they allege that bureaucrats involved in the preparation of *4AR* removed statements by scientists that highlighted climate risks, and that *4AR* therefore understates the risk of catastrophic warming. Thus David Wasdell (2007), an IPCC reviewer, writes that he was 'astounded at the alterations (to the final scientific draft of the full *4AR* report) that were imposed by government agents during

That's Where the Money Is

There have been repeated claims that this past year's hurricane activity was another sign of human-induced climate change. Everything from the heat wave in Paris to heavy snows in Buffalo has been blamed on people burning gasoline to fuel their cars, and coal and natural gas to heat, cool, and electrify their homes. Yet how can a barely discernible, one-degree increase in the recorded global mean temperature since the late 19th century possibly gain public acceptance as the source of recent weather catastrophes? And how can it translate into unlikely claims about future catastrophes?

The answer has much to do with misunderstanding the science of climate, plus a willingness to debase climate science into a triangle of alarmism. Ambiguous scientific statements about climate are hyped by those with a vested interest in alarm, thus raising the political stakes for policymakers who provide funds for more science research to feed more alarm to increase the political stakes. After all, who puts money into science—whether for AIDS or space or cli-

the final stage of review.' It obviously matters not whether bu-
reaucratic interference results in exaggerating the climate-
change risks or minimizing them; in either case, and as is now
agreed by both main sides to the global warming dispute, the
'consensus' advice tendered to governments by the IPCC is political and not scientific."[16]

To be fair to the IPCC, Lindzen points out that even scientific
groups that have been careful to avoid exaggeration and to identify
speculation as such have often had their work distorted by the
press. Exactly this happened when the White House asked the National Academy of Sciences, the country's premier scientific organization, to assemble a panel on climate change. The 11 members of
the panel, which included Lindzen, concluded that the science is

mate—where there is nothing really alarming? Indeed, the success
of climate alarmism can be counted in the increased federal spending on climate research from a few hundred million dollars pre-
1990 to $1.7 billion today. It can also be seen in heightened
spending on solar, wind, hydrogen, ethanol, and clean coal technologies, as well as on other energy-investment decisions.

But there is a more sinister side to this feeding frenzy. Scientists who dissent from the alarmism have seen their grant funds
disappear, their work derided, and themselves libeled as industry
stooges, scientific hacks, or worse. Consequently, lies about climate change gain credence even when they fly in the face of the
science that supposedly is their basis.

Alarm rather than genuine scientific curiosity, it appears, is essential to maintaining funding. And only the most senior scientists today can stand up against this alarmist gale and defy the iron
triangle of climate scientists, advocates, and policymakers.

Richard Lindzen, "Climate of Fear," April 12, 2006
Reprinted from The Wall Street Journal © 2006
Dow Jones & Company. All rights reserved.

far from settled: "Because there is considerable uncertainty in current understanding of how the climate system varies naturally and reacts to emissions of greenhouse gases and aerosols, current estimates of the magnitude of future warming should be regarded as tentative and subject to future adjustments (either upward or downward)."[17]

The press's spin? CNN, in language typical of other reportage, stated that it represented "a unanimous decision that global warming is real, is getting worse, and is due to man. There is no wiggle room."[18]

CHAPTER FIVE

Is It Warmer?

Vincent Gray, Syun-Ichi Akasofu, Robert Carter

Lindzen was just the fifth denier I had profiled. And yet already the headline case had taken some serious body blows, and the very notion of a consensus was beginning to seem questionable. Where to go from there? For my columns, my method really did not change very much. I kept looking for truly eminent—and interesting—exceptions to the science-is-settled claim and finding them pretty easily. One led to another and at any given moment I wrote the best story I had.

But when it came time to write this book, it seemed to me that the natural course was to follow the example my very first deniers gave me. To get a clear view of how vigorous the dissent is on the crucial issues, we need to look at the fundamental building blocks of the doomsayer case, step by step. We go backward to go forward. And we begin by going back to the very first question raised by Edward Wegman in my very first column: Is the Earth warmer?

Seems like a simple question, at first. But precisely as the hockey stick debate showed, that simple question immediately divides into more complex ones: Is the Earth warmer, compared to when? Did the warming start in the 18th century or the 20th? And most of all, is the amount of warming or the pace of warming so unusual in the history of climate as to demand a "special" explanation, like increased manmade CO_2 emissions?"

To begin with, though, let's take the simplest question. Has the Earth warmed in the 20th century? I did not expect to find any serious disagreement on that point—after all we have an instrumental record from weather stations around the world starting in the middle of the 19th century. And on that I was—almost right. There is not much dispute on this point. But there is just a little, and unfortunately it will probably persist for as long as the IPCC continues the high-handed and secretive behavior that drives those who question it to distraction, men like Dr. Vincent Gray.

Dr. Gray has become known as one of the most persistent, even vitriolic, critics of the IPCC. He has gone so far as to denounce the entire IPCC process as a "swindle."[1] That level of rhetoric might seem to disqualify him as a reliable witness, except that it is based on a long, intimate, and frustrating experience. You see, Dr. Gray is one of those "2,500 top scientists" from around the world whose good word the IPCC cites to assure us its reports can be trusted. That body has used Dr. Gray as an expert reviewer for many years. Vincent Gray, for his part, is one of the most active and energetic reviewers the panel has, logging roughly 1,900 comments on the IPCC's final draft of its most recent report alone.

The experience has made Dr. Gray an angry man, aghast at what he sees as an appalling absence of scientific rigor in the IPCC's review process. "Right from the beginning, I have had difficulty with this procedure. Penetrating questions often ended without any answer. Comments on the IPCC drafts were rejected without explanation, and attempts to pursue the matter were frustrated indefinitely."[2] Over time things got worse. "I have found increasing

THE CV OF A DENIER
Dr. Vincent Gray

Vincent Gray is a graduate of the University of Cambridge, with a Ph.D. in physical chemistry. He has published more than a hundred scientific papers and authored the book, *The Greenhouse Delusion: A Critique of "Climate Change 2001."* Dr. Gray has participated in all of the science reviews of the Intergovernmental Panel on Climate Change and in 2006 was a visiting scholar at the Beijing Climate Center.

opposition by them to providing explanations, until I have been forced to the conclusion that for significant parts of the work of the IPCC, the data collection and scientific methods employed are unsound. Resistance to all efforts to try and discuss or rectify these problems has convinced me that normal scientific procedures are not only rejected by the IPCC, but that this practice is endemic, and was part of the organization from the very beginning."[3]

So fundamental does he believe the problems to be that Dr. Gray now challenges even such basic assumptions of the IPCC process as the claim that we have accurate global temperature records, including for modern times. Gray notes, for instance, that temperature stations are not randomly distributed, with more than 90% on land, though 70% of the Earth's surface is covered by the oceans. Temperature stations are disproportionately located near cities and towns, which are heat sources, rather than out in the country. Moreover, many stations that were once in the country have had cities grow up around them, affecting temperature trends.

The IPCC has not ignored these charges, which in fact are

made by others as well as Gray. It says it is well aware of the difficulties in the raw data. The IPCC says it has taken appropriate statistical measures to clean and correct the data. That seems likely, given the obviousness of the objection.

Unfortunately, Gray's skepticism is supported by the IPCC's own secretive behavior. As Bob Carter, whom we met in a previous chapter, points out, IPCC practices make it impossible to settle such arguments. "The data used to construct the version of the global surface temperature used by the IPCC is not released to the public; the curve is therefore unreproduceable in the sense that it cannot be checked independently."[4] He does note that in some cases very public pressure has caused the IPCC to agree to release some data. The question remains: how much more thoroughly might the IPCC's temperature histories—and other assertions—be challenged if its data and methodologies were freely available to researchers? After all, in the case of the hockey stick, we saw a non-trivial example of this same problem: Mann's reluctance to share data and methods only heightened the suspicion of early skeptics like McIntyre. And then when the information became available, it turned out that Mann's methods were flawed. Strident and angry though he may be, under the circumstances one can't quite rule out Dr. Gray's suspicions about the temperature data.

Still, Gray's objections may be considered a sideshow to the main issues of "compared to when?" and "compared to what?"

If someone tells you that the Earth is warming and bases his claim on temperature changes in the last ten minutes, you will rightly dismiss this information as meaningless, even if it is perfectly true. If he then comes back and says, "now I have gone back a whole hour and it's clear the Earth is actually getting cooler," you will again dismiss his information.

So how long a trend is long enough to matter? That depends on the shape and frequency of past trends and cycles. If, bizarrely, the average temperature of the Earth had never varied over the previous hundred million years, the news that temperature suddenly rose a fraction of a degree from 1975 to 1998 would certainly be remarkable and might well be alarming. On the other hand, if we allow such short-term changes into evidence, then we have to let them all in. If temperature were then to level off or even decline over the next few years, this would be as much a "disproof" of global warming as the 1975 to 1998 rise is a "proof."

This is why the hockey stick was of such immense importance. By eliminating the Medieval Warming Period and making the 20th-century temperature increase look utterly unique in the course of a millennium, Mann and the IPCC got the world to sit up and take notice. Put the Medieval Warming Period back in, and the 20th-century warming looks much less interesting.

As Carter points out, the answer to the question: "Is global average temperature rising or falling? . . depends entirely on the chosen end-points of the data being considered."[5] For example, using the Greenland ice-core oxygen isotope data (proxy for local temperature), he explains, Greenland has gotten warmer over the past 16,000 years. It has also gotten warmer over the past 100 years. "Over intermediate time periods, however, cooling has occurred since 10,000 and 2,000 years ago, and temperature stasis characterizes both the last 700 years and (globally, from meteorological records) the last eight years [from 1998 through 2006]. Considering these facts, is the temperature in Greenland warming or cooling?"[6] (See fig. 1.)

Realistically, he argues, both the last eight years of zero warming (fig. 2) and the 100 years of warming preceding are "too short to carry statistical significance regarding long-term climate change."[7] No meaningful comparative judgments about climate change can "be made on the basis of the trivially short, 150-year-long thermometer surface temperature record, much less on the 27-year-long satellite tropospheric record."[8]

Nevertheless, since the global warming case is based largely on such short-term events, Carter thinks it interesting that Gray, using data from satellite-mounted microwave sounding units (MSU)

Whether the Earth is warming or cooling depends on the time scale chosen

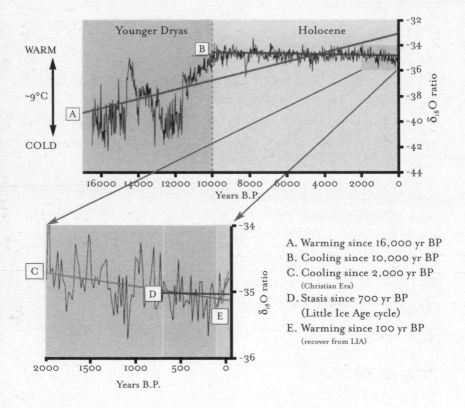

Figure 1. Climatic cycling over the last 16,000 years in Greenland. Carter notes that the temperature record used here is "as indicated by averaged 20-year oxygen isotope ratios from the GISP2 Greenland ice-core (after NSIDC User Services, 1997 and Davis and Bohling, 2001). Trend lines A–E all extend up to the end of the 20th century, fitted through the data for the last 16,000, 10,000, 2,000, 700, and 100 years, respectively. The trends are indicative of both warming and cooling, depending upon the chosen starting point, and all except E are statistically significant." *Adapted from* R. M. Carter, "The Myth of Dangerous Human-Caused Climate Change," The Australian Institute of Mining and Metallurgy New Leaders' Conference, Brisbane, QLD, May 2–3, 2007, p. 67. See: http://icecap.us/images/uploads/200705-03AusIMMcorrected.pdf.

Did the recent warming trend end in 1998?

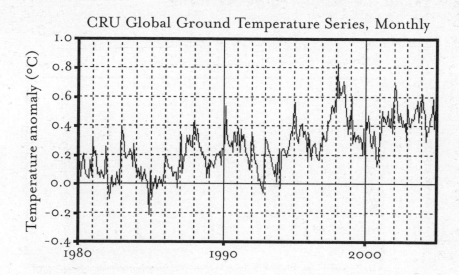

CRU Global Ground Temperature Series, Monthly

Figure 2. Carter argues that there has been no additional global warming since 1998. The figure shows: "Combined annual land, surface-air, and sea-surface global temperature anomalies (°C) for 1980–2005 relative to a 1961–1990-average baseline (data from Climate Research Unit, University of East Anglia). Though a warming of perhaps 0.3°C is recorded between 1980 and 1998 (a marked El Niño year), no warming has occurred in the seven subsequent years despite continued large increases in human-sourced atmospheric carbon dioxide." *Adapted from* R. M. Carter, "The Myth of Dangerous Human-Caused Climate Change," The Australian Institute of Mining and Metallurgy New Leaders' Conference, Brisbane, QLD, May 2–3, 2007, p. 67. See: http://icecap.us/images/uploads/200705-03AusIMMcorrected.pdf.

available since 1979, shows no warming in the troposphere since that time. "Of all these datasets, the MSU record is accepted to be the most accurate and globally representative. Once the effects of El Niño warmings and volcanic coolings are allowed for, this record shows no significant warming since its inception in 1979 (Gray, 2006) (fig. 3). This conclusion is robust. Though several other global temperature datasets exist, and though the MSU record has been subject to repeated corrections in interpretation,

Do events such as El Niño or volcanic activity explain late-20th-century warming?

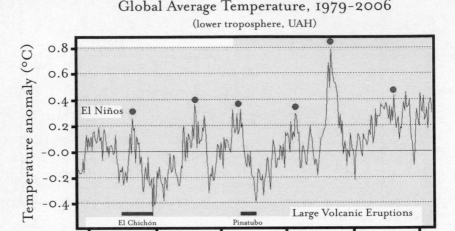

Global Average Temperature, 1979-2006
(lower troposphere, UAH)

Figure 3. Drawing on Gray, Carter cites evidence that even the 1979–1998 warming is explained by El Niño and other events: "Lower tropospheric temperature anomaly since 1979 as measured by satellite-mounted microwave sounding units (MSU; from http://vortex.nsstc.uah.edu/data/msu/t2lt/tltglhmam_5.2). When the warming effect of El Niños, and the cooling effect of the El Chichón and Pinatubo volcanic eruptions, are discounted, little if any greenhouse-forced warming is apparent for the last 25 years (Gray, 2006). Note also that these tropospheric measurements agree with the ground-based thermometer series (figure 2) in recording no significant warming since 1998, and probably none since 1982." *Adapted from* R. M. Carter, "The Myth of Dangerous Human-Caused Climate Change," The Australian Institute of Mining and Metallurgy New Leaders' Conference, Brisbane, QLD, May 2–3, 2007, p. 68. See: http://icecap.us/images/uploads/200705-03AusIMMcorrected.pdf.

none of the available datasets document significant recent greenhouse warming."[9]

Geophysicist Syun-Ichi Akasofu is founding director of the International Arctic Research Center of the University of Alaska Fairbanks. He has been a giant in Arctic research since his discovery in 1964 of the origin of storms in the aurora borealis. He has published more than 550 professional journal articles, has been the invited author of many encyclopedia articles, and was twice named one of the "1,000 Most Cited Scientists." He has been honored by the Royal Astronomical Society of London, the Japan Academy of Sciences, and the American Geophysical Union.

Here is a summary of his account of recent global temperature increases:

The Earth slowly but surely warmed over the course of the 20th century. The increase appears to have been about half a degree Celsius, close to the IPCC's own estimates.

The Earth slowly but surely warmed over the course of the 19th century too. The increase again seems to have been about half a degree Celsius.

THE CV OF A DENIER
Dr. Syun-Ichi Akasofu

Syun-Ichi Akasofu, founding director of the International Arctic Research Center of the University of Alaska Fairbanks, received his Ph.D. in geophysics in 1961. He has published more than 550 professional journal articles, authored or coauthored ten books, and has been the invited author of many encyclopedia articles. Twice named one of the "1,000 Most Cited Scientists," he has been honored by the Royal Astronomical Society of London, the Japan Academy of Sciences, and the American Geophysical Union. In 2003, he received The Order of the Sacred Treasure, Gold and Silver Star, from the emperor of Japan. As director of the UAF Geophysical Institute from 1986 to 1999, he helped establish it as a key research center in the Arctic. He also helped establish the Alaska Volcano Observatory.

England has been warming since 1660

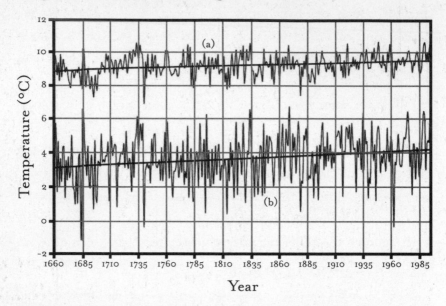

Figure 4. "The linear trends for the temperature of central England over the period 1660–1996 for (a) the annual data and (b) the winter months (December to February), show a marked warming. In both cases, this warming is significant, but although the temperature rise is greater in winter, this trend is less significant because the variance from year to year is correspondingly greater (L. D. Burroughs, 2001)." *Adapted from* Syun-Ichi-Akasofu, "Is the Earth Still Recovering from the 'Little Ice Age'?" International Arctic Research Center, University of Alaska Fairbanks, abstract, revised May 7, 2007, p. 6. See: http://www.iarc.uaf.edu/highlights/2007/akasofu_3_07Earth_recovering_from_LIA_R.pdf.

The Earth slowly but surely warmed over the course of the 18th century, too, global temperatures again increasing by about a half a degree Celsius.

Based on much more limited data, it appears that the Earth slowly but surely warmed over the course of the latter part of the 17th century as well, global temperatures again increasing at the rate of about a half a degree Celsius per century. Throughout these centuries, which followed the depths of the Little Ice Age, the rate of global warming has been fairly consistent.

"There is clearly a linear increase of temperature from about

1800 based on . . . ice-core data,"[10] says Akasofu. "Roughly the same linear change in temperatures extends back to the earliest recordings, going back to about 1660, even before the Industrial Revolution."[11]

So what is going on? Dr. Akasofu postulates one possible and startlingly straightforward explanation. The Little Ice Age is typically thought to have ended in 1900. Chances are good, he says, that it didn't. "The Earth may still be recovering from the Little Ice Age,"[12] he says, pointing to the consistent rate of warming over the centuries. If he is right, there is no need to invoke greenhouse gases or any manmade cause for the warming of the 20th century. If we are seeing the continuation of a natural trend, then there is nothing very special about recent warming.

Akasofu bases his argument for linear change in part on his own analysis of ice-core data from the Arctic island of Severnaya Zemlya. But he cites numerous other studies as well based on a variety of evidence, including work on other ice-core analyses. He summarizes this work in a now famous paper "Is the Earth Still Recovering from the 'Little Ice Age'? (The quotations and figures in this chapter are drawn from the May 7, 2007, revision of that paper. Just as this book was being readied for press, Akasofu released a revised version (February 11, 2008) of the paper, with much additional evidence and new figures. It is wonderfully easy reading even for laymen, and I strongly suggest reading the entire paper which can be found at http://people.iarc. uaf.edu/~sakasofu/ pdf/Earth_recovering_from_LIA_R.pdf.)

In that paper Akasofu notes, "There are several other supporting studies that suggest that there has been a linear change from about 1800 or earlier. For example, figures [4, 5, 6, 7, and 8] suggest a roughly linear change of temperature from the earliest recordings by Burroughs (2001), Tarand and Nordli (2001), and van Engelen et al. (2001). The trend lines and curves were drawn by the quoted authors, not by the present author."[13]

Numerous weather stations show a long-term gradual warming trend

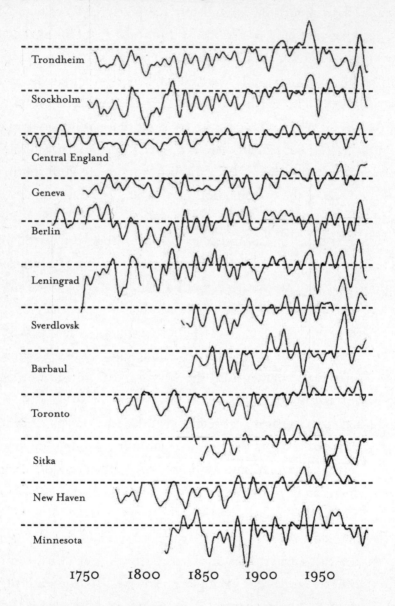

Figure 5. Temperature change at a number of stations in the world, after P. D. Jones and R. S. Braley, 1992. *Adapted from* Syun-Ichi-Akasofu, "Is the Earth Still Recovering from the 'Little Ice Age'?" International Arctic Research Center, University of Alaska Fairbanks, abstract, revised May 7, 2007, p. 6. See: http://www.iarc.uaf.edu/highlights/2007/akasofu_3_07/Earth_recovering_from_LIA_R.pdf.

Figures 4 through 8 all represent temperature studies. But other evidence exists as well. Glaciers are retreating and sea level is rising, just as global warming doomsayers point out. The problem, says Akasofu, is that they have been doing so in the same linear manner starting long before the 20th century. "There is further supporting evidence. . . . [T]he southern edge of sea ice in the Norwegian Sea has been continuously receding from about 1800 to the present. Further, there is a possibility that the present receding is related to an intense inflow of warm North

Ice break-up times in Tallinn, Estonia, have been trending earlier for centuries

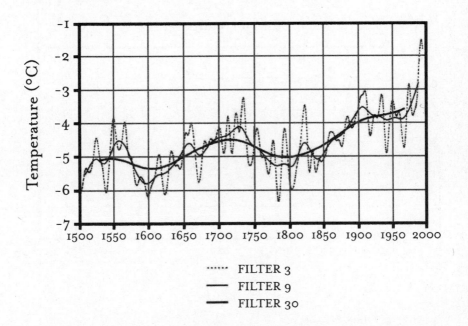

Figure 6. "Winter temperature (December–March) at Tallinn since 1500, which are based on ice break-up dates in Tallinn port. . . . After A. N. Tarand and P. O. Nordli, 2001." *Adapted from* Syun-Ichi-Akasofu, "Is the Earth Still Recovering from the 'Little Ice Age'?" International Arctic Research Center, University of Alaska Fairbanks, abstract, revised May 7, 2007, p. 7. See: http://www.iarc.uaf.edu/highlights/2007/akasofu_3_07/Earth_recovering_from_LIA_R.pdf.

Atlantic water (Polyakov et al., 2002); this phenomenon is known as the North Atlantic Oscillation (NAO), which is a natural phenomenon."[14] In addition, he points out that glaciers in Alaska and New Zealand "have been receding from the time of the earliest records. There are a large number of similar records from the European Alps and elsewhere (Grove, 1982). Therefore, it can be assumed that many glaciers advanced during the Little Ice Age and have been receding since then. Thus, the retreat is not something that began only in recent years."[15]

Summer temperatures in Tallinn show similar warming trend

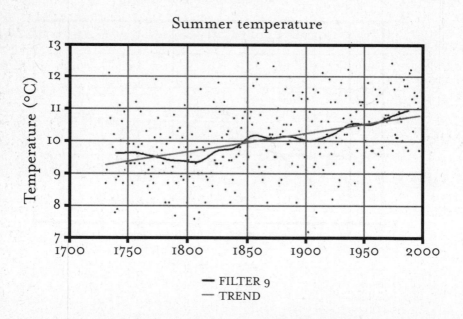

Figure 7. "Summer temperature (April to July) for Tallinn, which is based on ice break-up and rye harvest data and of instrumental observations....A trend line for the whole period is also shown." After (A. N. Tarand and P. O. Nordli, 2001). *Adapted from* Syun-Ichi-Akasofu, "Is the Earth Still Recovering from the 'Little Ice Age'?" International Arctic Research Center, University of Alaska Fairbanks, abstract, revised May 7, 2007, p. 7. See: http://www.iarc.uaf.edu/highlights/2007/akasofu_3_07/Earth_recovering_from_LIA_R.pdf.

Winter temperatures at DeBilt, Netherlands

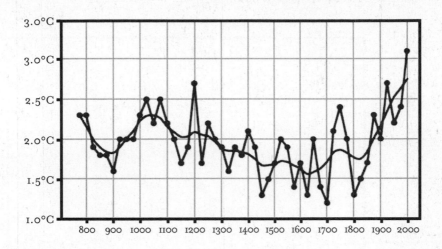

Figure 8. 25-year mean winter (DJF) temperature at De Bilt, Netherlands, after A. F.V. van Engelen, J. Buisman, and F. Ijnsen, 2001. *Adapted from* Syun-Ichi-Akasofu, "Is the Earth Still Recovering from the 'Little Ice Age'?" International Arctic Research Center, University of Alaska Fairbanks, abstract, revised May 7, 2007, p. 8. See: http://www.iarc.uaf.edu/highlights/2007/akasofu_3_07/Earth_recovering_from_LIA_R.pdf.

One especially dramatic example he cites is a 2007 recent study of sea level changes by Holgate. The study goes back only to 1904, so it sheds no light on climate changes in previous centuries. But it does show that in the 20th century, sea level rose significantly before 1940, in other words before the recent large increases in CO2 in the atmosphere that the doomsayers blame for global warming. "The sea level change should reflect the expected changes associated with the thermal expansion of seawater and glacier melting changes during the last half century that were warned in the IPCC Reports. Figure [9] shows that there is no clear indication of an accelerated increase of sea level after 1940. . . . In fact, comparing the slope between 1907–1960 and 1960–2000, there is even slightly less increase in the latter period. During the period of his study,

Gradual sea level rise appears to begin too early to be explained by human-caused global warming

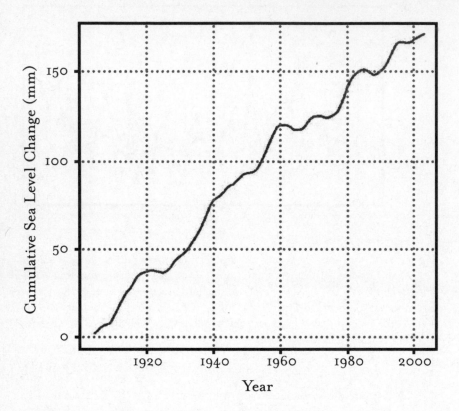

Figure 9. "Mean sea level record from the nine tide gauges over the period 1904–2003 based on the decadal trend values for 1907–1999. The sea level curve here is the integral of the rates (Holgate, 2007).[22]" *Adapted from* Syun-Ichi-Akasofu, "Is the Earth Still Recovering from the 'Little Ice Age'?" International Arctic Research Center, University of Alaska Fairbanks, abstract, revised May 7, 2007, p. 14. See: http://www.iarc.uaf.edu/highlights/2007/akasofu_3_07/Earth_recovering_from_LIA_R.pdf.

Holgate (2007) noted that the rate of sea level rise was about 1.7 mm/year"[16] (less than 7/100ths of an inch).

The implication? If temperature change and other related climate changes have been linear "without a distinct change of slope, from as early as 1800 or even earlier (about 1660, even before the Industrial

Revolution),"[17] then we are seeing a natural, not manmade phenomenon since "a rapid increase of CO_2 began only after 1940."[18]

Akasofu believes the slope of the temperature increase over the last several centuries "can roughly be estimated to be about 0.5°C/100 years"[19] based on the various studies he cites. Thus it is reasonable to attribute much of the warming of the 20th century to a continuation of whatever natural trend drove the increase in the immediate previous centuries. "There seems to be a roughly linear increase of the temperature of about 0.5°C/100 years (~1°F/100 years) from about 1800, or even much earlier, to the present."[20] That linear change of about half a degree every hundred years is very close to the 0.6 to 0.7° per century attributed to human causes by the IPCC. Since this long-standing linear trend predating the time when humans started pumping lots of CO_2 into the air is likely to be a natural change, and possibly an indication that the Earth is still recovering from the Little Ice Age, "[t]his trend should be subtracted from the temperature data during the last 100 years in estimating the manmade effect. . . . Thus, there is a possibility that only a fraction of the present warming trend may be attributed to the greenhouse effect resulting from human activities. This conclusion is contrary to the IPCC (2007) report (p. 10), which states that 'most' of the present warming is due to the greenhouse effect."[21]

He hastens to point out, however, that estimating the pace of the temperature change is not an easy or obvious task, especially given the need to rely on sketchy proxy data for early centuries. The natural change to be subtracted from the 20th-century change could be less than 0.5 degrees, or it could be more. But quite clearly a "significant" portion of the 20th-century warming was natural "contrary to the statement by the IPCC report (2007)."[22]

Akasofu notes that "it is well known that CO_2 causes the greenhouse effect, so that it is natural to hypothesize that CO_2 is one of the causes of the present warming trend."[23] Nevertheless, he is quite confident CO2 cannot be the primary cause. How can he be so confident on this point when he admits that temperature records before the late 19th century are not as reliable as we would like? That's a story for another day, or at least the next chapter.

Meanwhile, we'll leave you with a thought from the IPCC report of 2001, page 97. "The fact that the global mean temperature has increased since the late-19th century and that other trends have been observed does not necessarily mean that an anthropogenic effect on the climate system has been identified. Climate has always varied on all time scales, so the observed change may be natural. A more detailed analysis is required to provide evidence of a human impact."[24] Dr. Akasofu would certainly agree.

CHAPTER SIX

Looking For CO2

Tom Segalstad, Nir Shaviv,
Syun-Ichi Akasofu, Robert Carter et al.

The essence of the doomsayer case on global warming is that human activities dramatically increase the proportion of greenhouse gases in the atmosphere and that this increase is sufficient to increase global temperature by a potentially catastrophic amount.

Among all the deniers I have profiled, I have never encountered one who disputes that there is such a thing as a greenhouse effect, or that carbon dioxide is a greenhouse gas. It is an established physical property of carbon dioxide that, as Carter puts it in his literature survey, it "absorbs space-bound infrared radiation, thereby increasing the energy available at the Earth's surface for warming or increased evaporation."[1] The arguments are all about how powerful the effect is, especially when considered in combination with other factors, various feedback mechanisms both negative and positive, and other influences that might or might not overwhelm the effect of CO2.

And what that ends up meaning, really, is that once again the argument is about *when* as much as it is about *how*. The how of climate, I kept hearing, is extremely complicated and in crucial ways poorly understood. So researchers tend to fall back on *when*, that is, on correlations over time between, say, increases in atmospheric CO_2 and temperature. Thus just as the hockey stick was once Exhibit A for global warming, because it seemed to prove that 20th-century warming was a unique event on a time scale of the past thousand years, more recently Exhibit A has been that enormous 600,000-year graph of CO_2 vs. temperature that Al Gore stands in front of in his movie, *An Inconvenient Truth*, purporting to show that a rise in CO_2 always brings a rise in temperature.

Despite the 600,000-year graph, however, most of the argument for CO_2 as a driver of global warming involves much shorter time scales. Which brings us back to Dr. Akasofu.

Dr. Akasofu on CO_2

As an Arctic specialist, Akasofu notes that the Arctic acts like a magnifying glass through which one can examine climate change. When global average temperature changes moderately, Arctic temperature often changes dramatically.

Arctic data, for example, show a very large rise and then fall in temperature between 1920 and the early 1970s, while the global average data show this fluctuation as a minor blip. A second temperature fluctuation involves a rise after 1975. These fluctuations also look minor in warmer climes, but are very striking in the Arctic. (See figure 1.) "There occurred two major fluctuations, one between 1920 and 1975, and one after 1975. The Arctic data indicates that the two fluctuations in the global average data should not be ignored as minor fluctuations."[2]

"Once we 'take seriously' the Arctic data, then the very dramatic

Arctic temperature record correlates poorly with industrial CO2 emissions

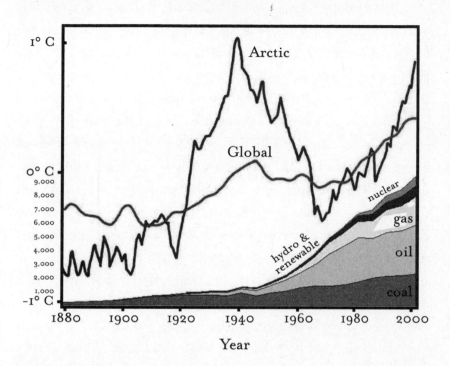

Figure 1. Arctic temperatures fell sharply for about 25 years just when CO2 emissions were accelerating. Moreover, the temperature rise from 1920 to 1940 was far more dramatic than the temperature rise from 1975 to 1998, at a time when hydrocarbon use was relatively small. Gray line: global average change from IPCC reports. Black line: based on data from stations along the coastline of the Arctic Ocean, after Polyakov et al., 2002. The figure shows also the amount of various sources of energy used during the last century: gas, oil, and coal all release CO2. *Adapted from* Syun-Ichi-Akasofu, "Is the Earth Still Recovering from the 'Little Ice Age'?" International Arctic Research Center, University of Alaska Fairbanks, abstract, revised May 7, 2007, p. 3. See: http://www.iarc.uaf.edu/highlights/2007/akasofu_3_07/Earth_recovering_from_LIA_R.pdf.

rise in temperature (more than 1.6 degrees) from 1920 to 1940, and the subsequent almost equally drastic fall in temperature from 1940 to the mid-1970s, cause serious problems for the CO2 thesis,"[3] says Akasofu. These dramatic changes in Arctic temperature don't correlate with changes in CO2 levels. Also they make the sharp rise in temperature from 1975 to 1998, which Mann made so much of, look like a normal fluctuation, since a very similar pattern occurs from 1920 to 1940. "Thus, the large fluctuation between 1920 and 1975 can be considered to be a natural change, until proven otherwise. . . . Contrary to the statement by the UN's Intergovernmental Panel on Climate Change in its 2007 report, it is not possible to say with any confidence that the rise after 1975 is mostly caused by the greenhouse effect."[4] Of course it is possible to argue that the 30-plus years of *rising* manmade CO2 emissions and *falling* temperature from 1940 to the mid-1970s are too short a time to prove anything. But then it becomes difficult to argue that the 20-plus years of rising CO2 and *rising* temperature, from the mid-1970s to 1998, prove anything either.

Akasofu sums up: "Indeed, there is so far no definitive proof that 'most' of the present warming is due to the greenhouse effect, as is stated in the recently published IPCC report (2007). In fact, the relationship between air temperature and CO2 is not simple. For example, the temperature had a cooling trend from 1940 to about 1975, in spite of the fact that atmospheric CO2 began to increase rapidly in about 1940. . . . It is not possible to determine the percentage contribution of the greenhouse effect that is a direct result of human activities, unless (and until) natural causes can be identified and subtracted from the present warming trend."[5]

Ironically, the IPCC's own climate-change models also point to carbon dioxide's irrelevance in climate change, says Akasofu. The Earth's warming is not uniform: different geographic regions are warming at different rates, while others are actually cooling. Dr. Akasofu asked the IPCC's Arctic group to apply its global-climate models to "hind-cast" the geographic distribution of the temperature change during the last half of the last century. ("Hind-casting" asks a model to produce results that match the known observations of the past—a model that can "predict" the past is more likely to be able to predict future conditions.)

To Akasofu's surprise, the model's results "predicted" dramatically different past temperatures than those obtained from actual readings, with no apparent relationship between the two. Initially, Dr. Akasofu thought the problem lay in the model. "However, this possibility is inconceivable, because the increase of CO2 measured in the past is correctly used in the hind-casting, and everything we know is included in the computation. It took a week or so for another possibility to dawn on us: if the warming and cooling is not caused by the greenhouse effect, the models will not show CO2-related warming and cooling."[7]

To examine that possibility, Dr. Akasofu checked to see if the magnified warming in the continental Arctic was still increasing in line with the ever-increasing amounts of CO2 entering the atmosphere. To his surprise, the continental Arctic had stopped its magnified warming and was now warming only at the same rate as the rest of the world. The upshot: in Akasofu's view the IPCC models tend to confirm that "[m]uch of the prominent continental Arctic warming during the last half of the last century is due to natural change."[8]

Dr. Tom Segalstad and the Mystery of the "Missing Sink" (of CO2)

Timing is also the central concern of another eminent scientist who disputes the CO2 as climate-driver hypothesis. Dr. Tom Segalstad is head of the Geological Museum at the University of Oslo and formerly an expert reviewer with the IPCC. He has also served as head of the Mineralogical-Geological Museum at the University of Oslo, director of the Natural History Museums and Botanical Garden of the University of Oslo, and program chairman for mineralogy/petrology/geochemistry at the University of Oslo.

As with any atmospheric gas, atmospheric CO_2 has both "sources," which add CO_2 to the air, and "sinks," which absorb CO_2 from the air. Your car is a source, and so are you. Green plants are a sink. The oceans swing both ways: they have enormous capacity to absorb CO_2, but sometimes they give it up.

The IPCC, in making the case that human activities will massively increase CO_2 and drive catastrophic temperature increase, claims that manmade CO_2 stays in the atmosphere for 50, 100, or even 200 years. This unprecedented buildup of CO_2 then traps heat that would otherwise escape our atmosphere, threatening us all. This assumption, that added CO_2 stays in the atmosphere for many decades, is crucial to the doomsayer thesis, because it justifies projections that human activity will enormously increase CO_2 levels in the atmosphere.

"This is nonsense,"[9] says Segalstad. It reflects the paucity of geologic knowledge among IPCC scientists, dangerous since geologic processes ultimately determine the level of atmospheric CO_2. "The IPCC needs a lesson in geology to avoid making fundamental mistakes,"[10] he says, adding that "most leading geologists, throughout the world, know that the IPCC's view of Earth processes are implausible if not impossible."[11]

Until recently, the world of science was near unanimous that CO_2 couldn't stay in the atmosphere for more than about five to ten years because of the oceans' near-limitless ability to absorb CO_2.

"This time period has been established by measurements based on natural carbon-14 and also from readings of carbon-14 from nuclear weapons testing. It has been established by radon-222 measurements. It has been established by measurements of the solubility of atmospheric gases in the oceans. It has been established by comparing the isotope mass balance. It has been established through other mechanisms, too, and over many decades.

And by many scientists in many disciplines,"[12] says Prof. Segalstad, whose work has often relied upon such measurements.

With the advent of IPCC-influenced science, the length of time that carbon stays in the atmosphere became controversial. As Segalstad tells the story, the claim that CO2 levels in the atmosphere had increased substantially in the 20th century because of human activity has been elevated to the status of dogma, an unquestionable axiom that governs all subsequent hypotheses. As Segalstad points out, there has been all along a very vigorous debate about this claim, with eminent scientists such as Jawarowski and others (see chapter seven) arguing that CO2 levels fluctuate naturally and that levels as high as today's were not uncommon in the 19th century.

Once the notion of rising CO2 levels is accepted as dogma, then even well-established science must be rewritten to accommodate it. Studies of the "lifetime" or "residence" of CO2 in the atmosphere, before it is absorbed into the ocean, have overwhelmingly given a lifetime of from five to ten years, perhaps as long as 12 years. If this traditional assumption of short residence times is correct, then humans have not been pumping out CO2 fast enough to account for the claimed increase of CO2 in the atmosphere during the 20th century. This in turn yields two possibilities: either something other than human activity is driving the alleged secular increase in CO2 levels in this century, or there has been no dramatic secular increase, for instance, because the measuring methodologies are flawed. (Again, see chapter seven.)

Either is a reasonable possibility, worthy of investigation. Alas, as Segalstad points out, both contradict the central dogma of climate-change science that CO2 levels have been rising and humans are the cause. To preserve the dogma, he argues, the global warming movement needed new carbon cycle models that, practically speaking, start from the assumption that the central dogma is correct. If the dogma requires long residence times, then the models shall produce them. These models are now pillars of the global warming case as presented by the UN.

More than three-dozen studies contradict UN-IPCC's claim that increased CO2 persists in the atmosphere long term

AUTHORS [publication year]	RESIDENCE TIME (years)
IPCC estimate [2007]	50-200

BASED ON NATURAL CARBON-14

Craig [1957]	7 +/- 3
Revelle & Suess [1957]	7
Arnold & Anderson [1957]	10
including living and dead biosphere (Siegenthaler, 1989)	4-9
Craig [1958]	7 +/- 5
Bolin & Eriksson [1959]	5
Broecker [1963], *recalculated by* Broecker & Peng [1974]	8
Craig [1963]	5-15
Keeling [1973b]	7
Broecker [1974]	9.2
Oeschger et al. [1975]	6-9
Keeling [1979]	7.53
Peng et al. [1979]	7.6 (5.5-9.4)
Siegenthaler et al. [1980]	7.5
Lal & Suess [1983]	3-25
Siegenthaler [1983]	7.9-10.6
Kratz et al. [1983]	6.7

BASED ON "SUESS EFFECT"*

Ferguson [1958]	2 (1-8)
Bacastow & Keeling [1973]	6.3-7.0

BASED ON BOMB CARBON-14

Bien & Suess [1967]	>10
Münnich & Roether [1967]	5.4
Nydal [1968]	5-10
Young & Fairhall [1968]	4-6

Rafter & O'Brian [1970] 12
Machta (1972) 2
Broecker et al. [1980a] 6.2–8.8
Stuiver [1980] 6.8
Quay & Stuiver [1980] 7.5
Delibrias [1980] 6.0
Druffel & Suess [1983] 12.5
Siegenthaler [1983] 6.99–7.54

BASED ON RADON-222
Broecker & Peng [1974] 8
Peng et al. [1979] 7.8–13.2
Peng et al. [1983] 8.4

BASED ON SOLUBILITY DATA
Murray (1992) 5.4
**BASED ON CARBON-13/
CARBON-12 MASS BALANCE**
Segalstad (1992) 5.4

*Caused by the addition of old carbon-14-free CO_2 from combustion of fossil fuels

Figure 2. Atmospheric residence time of CO_2, mainly based on the compilation by Sundquist, 1985. Bracketed references are from Sundquist. *Adapted from* Tom V. Segalstad, "Carbon Cycle Modeling and the Residence Time of Natural and Anthropogenic Atmospheric CO_2: On the Construction of the 'Greenhouse Effect Global Warming' Dogma," 1997, from "Global Warming: The Continuing Debate," European Science and Environment Forum (ESEF), (Cambridge, England [ISBN 0-9527734-2-2] 1998), pp. 184–219. See: http://folk.uio.no/tomvs/esef/ESEF3VO2.pdf.

Amazingly, the hypothetical results from climate models have trumped the real-world measurements of carbon dioxide's longevity in the atmosphere. Those who claim that CO2 lasts decades or centuries have no such measurements or other physical evidence to support their claims, says Segalstad. Neither have they demonstrated that the quite various forms of measurement that support the traditional five-to-ten-year view are wrong.

"They don't even try,"[13] says Prof. Segalstad. "They simply dismiss evidence that is, for all intents and purposes, irrefutable. Instead, they substitute their faith, constructing a kind of science fiction or fantasy world in the process."[14]

In the real world, as measurable by science, CO_2 in the atmosphere and in the ocean reach a stable balance when the oceans contain 50 times as much CO_2 as the atmosphere. "The IPCC postulates an atmospheric doubling of CO_2, meaning that the oceans would need to receive 50 times more CO_2 to obtain chemical equilibrium,"[15] explains Prof. Segalstad. "This total of 51 times the present amount of carbon in atmospheric CO_2 exceeds the known reserves of fossil carbon—it represents more carbon than exists in all the coal, gas, and oil that we can exploit anywhere in the world."[16]

THE CV OF A DENIER
Professor Tom V. Segalstad

Professor Tom V. Segalstad is head of the Geological Museum within the Natural History Museums of the University of Oslo. Formerly, he was head of the Mineralogical-Geological Museum at the University of Oslo, director of the Natural History Museums and Botanical Garden of the University of Oslo, and program chairman for mineralogy/petrology/geochemistry at the University of Oslo. His research projects include geological mapping in Norway, Svalbard (Arctic), Sweden, and Iceland and have involved geochemistry, volcanology, metallogenesis (how mineral and ore deposits form), and magmatic petrogenesis (how magmatic rocks form). He was an expert reviewer to the UN's Intergovernmental Panel of Climate Change's *Third Assessment Report*.

Also in the real world, Prof. Segalstad's isotope mass balance calculations—a standard technique in science—show that if CO2 in the atmosphere had a lifetime of 50 to 200 years, as claimed by IPCC scientists, the atmosphere would necessarily have half of its current CO2 mass. Because this is a nonsensical outcome, the IPCC model postulates that half of the CO2 must be hiding somewhere, in "a missing sink."[17] Many studies have sought this missing sink—a Holy Grail of climate-science research—without success. To the contrary, says Segalstad, the models searching for the missing sink actually turn up additional sources, compounding the problem.

"When trying to find this 'missing sink' in the biosphere, carbon cycle modelling has shown that deforestation must have contributed a large amount of CO2 to the atmosphere. So instead of finding the 'missing sink' in the terrestrial biosphere, they find another CO2 source! This makes the 'missing sink' problem yet more severe."[18] An error of this stunning magnitude equating to "about 3 giga-tonnes of carbon annually not explained by a model," says Segalstad, "would normally lead to complete rejection of the model and its hypothesis."[19]

To sum up, says Segalstad, in order to support the central dogma of a large, secular human-caused increase in atmospheric CO2, climate-change scientists have launched "a search for a mythical CO2 sink to explain an immeasurable CO2 lifetime to fit a hypothetical CO2 computer model that purports to show that an impossible amount of fossil-fuel burning is heating the atmosphere."[20]

"It is all a fiction."[21]

In both my columns and this book, I have focused on quality, not quantity. But in many cases, and certainly with Segalstad, the story of one man's dissent readily reveals the teeming and contentious debate in the background. Segalstad's 1997 paper on carbon cycle modeling subtitled "On the Construction of the

'Greenhouse Effect Global Warming' Dogma," is relatively readable for laymen. (That is a recommendation, if not actually an assignment.) A survey of the relevant research, it powerfully suggests that far from being alone in his objections to the UN's CO_2 models, Segalstad and his colleagues have quite a good claim to the title of "mainstream" for their own allegedly heretical views.

Nir Shaviv and the Strange Case of Incrimination by Elimination

If such powerful objections exist to the CO_2 thesis, why have so many scientists endorsed it so readily?

Astrophysicist Nir Shaviv, one of Israel's top young scientists, gives an answer out of his own experience. He describes the logic that led him—and so many others—to conclude that SUVs, coal, plants, and other things manmade cause global warming.

Step One: scientists for decades have postulated that increases in carbon dioxide and other gases could lead to a greenhouse effect.

Step Two: as if on cue, the temperature rose over the course of the 20th century while greenhouse gases proliferated due to human activities.

Step Three: apparently no other mechanism explained the warming; without another candidate, greenhouses gases won the prize.

Dr. Shaviv, a prolific researcher who has made a name for himself assessing the movements of 2-billion-year-old meteorites, no longer accepts this logic or subscribes to these views: "Like many others, I was personally sure that CO_2 is the bad culprit in the story of global warming. But after carefully digging into the evidence, I realized that things are far more complicated than the story sold to us by many climate scientists or the stories regurgitated by the media."[22] He does not dismiss CO_2 as a contributing factor but has come to believe that it plays only a subordinate role and that the evidence for even this subordinate role, though "incriminating," is "circumstantial."[23] As he dug more deeply into the issue, Shaviv realized that according to the IPCC's own findings, man's role is so uncertain that it is even possible that we have been cooling, not warming, the Earth. Although science understands how CO_2 in and of itself within a simple system can magnify the Sun's warming of the Earth, Shaviv argues that we have no such clear theory about how the interaction of various warming and cooling factors affect the net temperature change. We understand CO_2 relatively well, but many other factors relatively poorly. A climate scientist focusing on CO_2 is a bit like the man who lost his keys half a block away but looks under the light post because that is where he can see.

Dr. Shaviv writes: "If we wish to assess theoretically how much is the anthropogenic contribution to 20th-century warming, we have to address two questions: How much is the anthropogenic contribution to the changed radiation budget, and how changes in the radiation budget affect the global temperature."[24] Very roughly the contribution to the "radiation budget" means how much a given factor magnifies or reduces the warming effect of the Sun. "For example, doubling the amount of CO_2 in the atmosphere changes the radiative budget by" enough that it is "as if the sun was 1.7% brighter."[25]

Naturally, as Shaviv explains, the IPCC reports have tried to "summarize the effects of all the drivers"[26] of temperature change. This is displayed in the "famous forcing graph"[27] from its 2001 report. (See figure 3.) Unfortunately, he says, there are significant uncertainties in these estimates, including a "large uncertainty in an anthropogenic contribution called the indirect aerosol effect. This effect arises from the fact that increased amounts of small particles in the atmosphere will alter the characteristics of clouds. This is best seen downstream of chimney stacks or in marine clouds in the form of ship tracks. Since cloud formation, and in particular the characteristics of clouds, is not well understood, the indirect aerosol effect is highly uncertain."[28] And of course burning hydrocarbons typically releases both CO_2 and aerosols into the air, so there may be some canceling effect. In addition, Shaviv argues, the IPCC numbers seriously underestimate how much certain changes in the Sun's behavior affect global temperature. (In this he is joined by a growing number of space scientists as we will see in chapters nine and ten.)

As a result of these uncertainties, Shaviv argues not only that we do not know how powerfully man is affecting the "radiation budget," we don't even know whether our effect is positive or negative. "[W]e may have actually been cooling Earth (though less likely than warming). It is for this reason that in the 1970s, concerns were raised that humanity is cooling the global temperature. The global temperature appeared to drop between the 1940s and 1970s, and some thought that anthropogenic aerosols could be the cause of the observed global cooling, and that we may be triggering a new ice age."[29]

Not knowing exactly the contribution of human activity to the radiative budget is only the first problem. We also "need to know the global-climate sensitivity. . . . The reason is that as the temperature changes, other variables affecting the temperature change as well. For example, increasing the radiation budget increases the temperature. This will increase the amount of water vapor in the atmosphere. However, water vapor is a great GHG [Green House Gas]. So this will tend to increase the temperature further, thus giving rise to a positive feedback, which increases the sensitivity. On the other hand, the larger amounts of water vapor in the atmosphere imply more cloud cover. Since clouds have a net tendency to cool, this will counter the increase in temperature, thus giving rise to a negative feedback that decreases the sensitivity."[30]

Looking under the light bulb

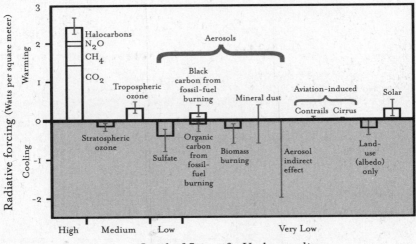

The global mean radiative forcing of the climate system for the year 2000, relative to 1750

Figure 3. Naturally, climate issues we understand relatively well dominate our explanations of global warming. But this "natural" prejudice also means we ignore or guesstimate factors that may be enormously important, such as clouds, land use, and the Sun. Shaviv comments: "If one adds [the IPCC's] numbers . . . one finds an anthropogenic forcing of: $0.8 \pm 1.3 \, W/m^2$ . . . In other words, the large uncertainty in the indirect aerosol effects implies that the sign of the anthropogenic contribution is unknown!" Figure *adapted from* the IPCC *Third Assessment Report* as cited in Nir J. Shaviv, "Carbon Dioxide or Solar Forcing," ScienceBits.com, http://www.sciencebits.com/CO2orSolar.

This presents grave problems for climate models. "The problem with numerical simulations of climate is that the feedbacks, especially those pertaining to cloud cover, are very poorly understood."[31] So great are the approximations involved that even on the models employed by the IPCC "the temperature change associated with doubled CO2 is not known to within a factor of 3!"[32]

Where does this leave us? Right back with Shaviv's chain of circumstantial evidence. We know that greenhouse gases are, in

THE CV OF A DENIER
Professor Nir Shaviv

Nir Shaviv, an associate professor at the Racah Institute of Physics at the Hebrew University of Jerusalem, received his doctorate from the Israel Institute of Technology in 1996. Since then, he has authored or coauthored some three-dozen peer-reviewed studies and presented papers at some two-dozen conferences. The Smithsonian/NASA Astrophysics Data System credits his works with a total of 613 citations. Among his prizes is the Beatrice Tremaine Award from the Canadian Institute for Theoretical Astrophysics at the University of Toronto.

themselves, a warming factor. We know that (on some time scale) the Earth is warming. And—that's it. Such circumstantial evidence might be a fine basis on which to justify reducing greenhouse gases, he adds, "without other suspects."[33] However, Dr. Shaviv not only believes there are credible other suspects, he believes that at least one provides a superior explanation for the 20th-century's warming.

"Solar activity can explain a large part of the 20th-century global warming,"[34] he states. Shaviv believes the mechanism linking solar activity to temperature changes is the Sun's impact on something called "cosmic ray flux," which will be explained more fully in chapter nine. He argues that in the past decade so much evidence has accumulated for this mechanism that it is "unlikely that it [the solar-climate link] does not exist."[35] Shaviv's own research into how the solar system's progress through the galaxy affects cosmic ray flux and how those changes correlate with climate change have made a significant contribution to this growing body of work, and have led him to conclude that natural solar processes may account for 80% of the global warming in the 20th century. Curiously, Shaviv attributes just about as much of global warming to natural causes as Akasofu.

Nevertheless, Dr. Shaviv maintains fossil fuels should be controlled not because of their adverse affects on climate but to curb pollution. "I am therefore in favor of developing cheap alternatives such as solar power, wind, and of course fusion reactors (converting deuterium into helium), which we should have in a few decades, but this is an altogether different issue."[36]

Carter . . . and Company

As with Segalstad's survey, Carter's, cited in previous chapters, shows that the ranks of those whose scientific work contradicts the "central dogma" are full indeed. Once again, Carter's survey is especially useful, and we quote it here at length. His sources are briefly referenced in parentheses. Full citations can be found in his paper "The Myth of Dangerous Human-Caused Climate Change," which can be found at a variety of URLs including www.lavoisier.com.au/papers/Conf2007/Carter2007.pdf.[37]

- High CO2 levels are not new: "Carbon dioxide levels during the recent past are very low by the standards of earlier geological history, for planetary carbon dioxide values have declined from around 1,000 ppm in the early Cenozoic, 60 million years ago (Lowenstein and Demicco, 2006)."[38]

- CO2 levels have fluctuated in the past, including fluctuations on time scales measured in decades or centuries, suggesting 20th-century fluctuations are not unusual. In the relatively recent, but pre-industrial past, CO2 levels have spiked as high as today's level of about 380 ppm. "Independent evidence from fossil plant stomata indicates that carbon dioxide levels during the Holocene [the era beginning at the end of the last great Ice Age, about 11,000 years ago] were variable on a decadal–centennial basis . . . and reached at least the present day (post-industrial) value of 380 ppm (Leuschner et al., 1996; Wagner, Aaby, and Visscher, 2002; Kouwenberg et al., 2005) . . . More support for decadal fluctuations of carbon dioxide comes from the compilation and summary of 90,000 historical atmosphere analyses back to the mid-19th century by Beck (2007)."[39]

- High CO_2 levels in the range of those predicted by the doomsayer thesis are not dangerous: "Prima facie, therefore, there is no reason to assume that atmospheric carbon dioxide levels of 500–1,000 ppm are dangerous, or that such levels would have dramatically adverse ecological effects. Rather, increasing atmospheric carbon dioxide over this range is mostly beneficial (Idso, 2001; and many papers listed at the Web site CO2 Science). . . . Increasing carbon dioxide in the range of about 200–1,000 ppm has repeatedly been shown to be beneficial for plant growth, and to increase plants' efficiency of water use (Eamus, 1996; Saxe, Ellsworth, and Heath, 1998; Robinson et al., 1998)."[40]

- As CO_2 levels rise it takes more and more CO_2 to produce additional temperature increases: "[T]he relationship between increasing carbon dioxide and increasing temperature is logarithmic, which lessens the forcing effect of each successive increment of carbon dioxide."[41]

- Most of the CO_2 warming should already have happened, suggesting future CO_2-driven warming would be trivial: "[I]n increasing from perhaps 280 ppm in pre-industrial times to 380 ppm now, carbon dioxide should already have produced 75 percent of the theoretical warming of ~1°C that would be caused by a doubling to 560 ppm (Lindzen, 2006); as we move from 380 to 560 ppm, at most a trivial few tenths of a degree of warming remain in the system. Claims of greater warming, such as those of the IPCC (2001), are based upon arbitrary adjustments to the lambda value in the Stefan-Boltzmann equation, and untested assumptions about positive feedbacks from water vapor."[42]

- Like Shaviv, Carter makes the point that despite Gore's famous 600,000-year chart, changes in temperature often appear to precede changes in CO2 levels and often do not correlate at all, making the cause-and-effect relationship ambiguous: "[T]he ice-core data show conclusively that, during natural climate cycling, changes in temperature precede changes in carbon dioxide by an average 800 years or so (Fischer et al., 1999; Indermuhle et al., 2000; Mudelsee, 2001; Caillon et al., 2003); similarly, temperature change precedes carbon dioxide change, in this case by five months, during annual seasonal cycling (Kuo, Lindberg, and Thomson, 1990). . . . Boucot, Xu, and Scotese (2004) have shown that over the Phanerozoic little relationship exists between the atmospheric concentration of carbon dioxide and necessary warming, including that extensive glaciation occurred between 444 and 353 million years ago when atmospheric carbon dioxide was up to 17 times higher than today (Chumakov, 2004)."[43]

There are obviously significant numbers of well-qualified people on the other side of the debate. But, as ever, our job is not to settle which side is right but simply to demonstrate that there is a debate, a vigorous and serious one. As Carter puts it, in "summary, there is almost universal agreement that significant carbon dioxide increases—human-caused or otherwise—will cause gentle planetary warming. But scientific opinion remains strongly divided as to how great a warming would accompany a real-world doubling and whether any such warming will on balance be beneficial or harmful."[44]

What about Al Gore's dramatic graph showing CO2 and temperature move together?

Adapted from "The Inconvenient Truth About the Ice Core Carbon Dioxide Temperature Correlations," by Nir Shaviv, *http://www.sciencebits.com/*

This famous graph doesn't prove that CO2 has any effect on the global temperature. All it says is that there is some equilibrium between dissolved CO2 and atmospheric CO2, an equilibrium that depends on the temperature. Of course, the temperature itself can depend on a dozen different factors, including CO2, but just the CO2/temperature correlation by itself doesn't tell you the strength of the CO2 link. It doesn't even tell you the sign.

Think of a closed Coke bottle. It has Coke with dissolved CO2 and it has air with gaseous CO2. Just like Earth, most of the CO2 is in the dissolved form. If you warm the Coke bottle, the Coke cannot hold as much CO2, so it releases a little and increases the partial pressure of the gaseous CO2, enough to force the rest of the dissolved CO2 to stay dissolved. Since there is much more dissolved CO2 than gaseous CO2, the amount released from the Coke is relatively small.

Of course, the comparison can go only so far. The mechanisms governing CO2 in the oceans are much more complicated: the equilibrium depends on the amount of biological activity, on the complicated chemical reactions in the oceans,

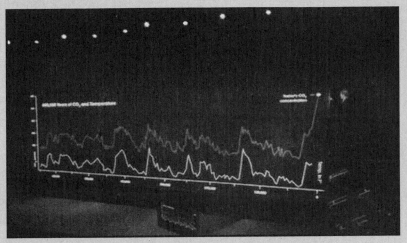

From *An Inconvenient Truth*, Paramount Classics and Participant Productions, featuring Al Gore, directed by Davis Guggenheim, produced by Laurie David, Lawrence Bender, and Scott Z. Burns, 2006.

and many more interactions I am probably not aware of. The bottom line is that the equilibrium is quite complicated to calculate.

The main evidence proving that CO2 does not control the climate, but at most can play a second fiddle by just amplifying the variations already present, is that of lags. In all cases where there is a good enough resolution, one finds that the CO2 lags behind the temperature by typically several hundred to a thousand years. Thus, the basic climate driver that controls the temperature cannot be that of CO2. That driver, whatever it is, affects the climate equilibrium, and the temperature changes accordingly. Once the oceans adjust (on a time scale of decades to centuries), the CO2 equilibrium changes as well. The changed CO2 can further affect the temperature, but the CO2/temperature correlation cannot be used to say almost anything about the strength of this link.

CHAPTER SEVEN

The Ice-core Man and the Great CO2 Cover-up

Zbigniew Jaworowski

O nce upon a time, and for millennia before then, carbon dioxide levels in the atmosphere were low and stable. Then came the Industrial Revolution and CO2 levels began to rise. The more man industrialized, the more that CO2—and the temperature—rose. In the last half century, with industrialization at unprecedented levels, CO2 reached levels unprecedented in human history. This is the story of global warming.

This story is a fable, says Zbigniew Jaworowski, past chairman of the Scientific Committee of the Central Laboratory for Radiological Protection in Warsaw, past chairman of the United Nations Scientific Committee on the Effects of Atomic Radiation, and a participant or chairman of some 20 Advisory Groups of the International Atomic Energy Agency and the United Nations Environmental Program.

Dr. Jaworowski agrees that CO2 levels rose in the last half century. Starting in 1958, direct real-time measurements of CO2 have been systematically taken at a state-of-the-art measuring station in Mauna Loa, Hawaii. These measurements, considered the world's most reliable, are a good basis for science by bodies like the UN's Intergovernmental Panel on Climate Change, the agency that is coordinating the worldwide effort to stop global warming.

But the UN does not rely on direct real-time measurements for the period prior to 1958. "The IPCC relies on ice-core data—on air that has been trapped for hundreds or thousands of years deep below the surface,"[1] Dr. Jaworowski explains. "These ice cores are a foundation of the global warming hypothesis, but the foundation is groundless—the IPCC has based its global warming hypothesis on arbitrary assumptions and these assumptions, it is now clear, are false."[2]

The IPCC works from the assumption that ice precisely preserves the ancient air, allowing for a precise reconstruction of the ancient atmosphere. For this to be true, no component of the trapped air can escape from the ice. Neither can the ice ever become liquid. Neither can the various atmospheric gases trapped in air bubbles in the ice ever combine or separate. And for the trapped gases to remain in proportion, they would also all need to react the same way to pressure from the weight of the ice above.

This perfectly closed system, frozen in time, is a fantasy. "Liquid water is common in polar snow and ice, even at temperatures as low as -72°C,"[3] Dr. Jaworowski explains, "and we also know that in cold water, CO_2 is 70 times more soluble than nitrogen and 30 times more soluble than oxygen, guaranteeing that the proportions of the various gases that remain in the trapped, ancient air will change. Moreover, under the extreme pressure that deep ice is subjected to—320 bars, or more than 300 times normal atmospheric pressure—high levels of CO_2 get squeezed out of ancient air."[4] This squeezing of CO_2 begins to happen at a depth of about 70 meters below the surface of the ice. At that depth, pressure from the weight of the ice above starts to crystallize the CO_2 in the air bubbles into solids. At 600 meters the oxygen begins to crystallize and at 1,000 meters, when the nitrogen begins to crystallize, the air bubbles start to disappear.

Because of these various properties of ice, one would expect that, over many centuries, ice that started off with high levels of CO_2 would become depleted, leaving behind a fairly uniform and base level of CO_2. In fact, this is exactly what the deep ice cores show.

"According to the ice-core samples, CO_2 levels vary little over time,"[5] Dr. Jaworowski states. "The ice-core data from the Taylor Dome in Antarctica shows almost no change in the level of atmospheric CO_2 over the last 7,000 to 8,000 years—it varied between 260 parts per million and 264 parts per million."[6]

If it is true that CO_2 has escaped from long-lived ice, the ice will not reveal the CO_2 levels that have existed in the history of the atmosphere. Other proxies for CO_2 in ancient air would be expected to give very different readings. That, says Dr. Jaworowski, is exactly what we find. "Other indicators of past CO_2 levels, such as fossil leaf stomata, show that CO_2 levels over the past 7,000 to 8,000 years varied by more than 50 parts per million, between 270 and 326 parts per million,"[7] the latter number comparable to levels today. We also know that there have been great fluctuations in temperature over that time period, "the Little Ice Age just 750 to 130 years ago, for example. If the ice-core record was reliable, and CO_2 levels reflected temperatures, why wouldn't the ice-core data have shown CO_2 levels to fall during the Little Ice Age or, for that matter, increase during the Medieval Warming around the year 1000?"[8]

Ice-core data from more recent times, shallower than 200 meters, cover the period since the Industrial Revolution. But shallow ice cores pose a challenge for scientists determined to demonstrate that CO_2 levels have risen steadily with industrialization, Dr. Jaworowski explained in his March 19, 2004, statement to the U.S.

THE CV OF A DENIER
Zbigniew Jaworowski

Zbigniew Jaworowski is senior scientific advisor of the Scientific Council of the Central Laboratory for Radiological Protection in Warsaw, Poland, where he has held various posts since 1973, including chairman from 1995 to 2007. He was a principal investigator of three research projects of the U.S. Environmental Protection Agency and of four research projects of the International Atomic Energy Agency. The author of four books and nearly 300 scientific papers, he has held posts with the Center d'Etude Nucleaires near Paris; the Biophysical Group of the Institute of Physics, University of Oslo; the Norwegian Polar Research Institute; and the National Institute for Polar Research in Tokyo. He has represented Poland on the United Nations Scientific Committee on the Effects of Atomic Radiation (UNSCEAR) continuously since 1973 and was its chair from 1980 to 1982.

Senate Committee on Commerce, Science, and Transportation.[9] Ice cores taken at Siple, Antarctica, provide one such example.

As might be expected, the Siple ice cores show an inverse correlation between CO_2 concentrations and load pressure: the deeper the ice, the greater the pressure and thus the lower the concentration of CO_2. (See figures 1A and B.) The ice deposited in 1890, for example, at a depth of 68 meters where the pressure is 5 bars, has a CO_2 concentration of 328 ppmv (parts per million volume), compared to a concentration of 280 ppmv in 1660 at a depth of 200 meters, where the pressure is 15 bars.

The 1890 reading of 328 ppmv, however, conflicts with IPCC

models that claim CO2 levels have been steadily increasing since the Industrial Revolution in the 1700s. According to the models, CO2 concentrations in 1890 must have been about 290 ppmv, not 328 ppmv, which the measurements at Mauna Loa, Hawaii, show weren't reached until 1973—83 years later.

Figure 1A below shows the "problem" very clearly. The Siple data appear to follow the shape of the Mauna Loa curve quite nicely. And the early low levels of CO2 reflected by the Siple data fit with the theory that CO2 levels were lower until industrialization.

How to make the data fit the theory

Figure 1A and B. Concentration of CO2 in air bubbles from the pre-industrial ice from Siple, Antarctica, (open squares), and in the 1958–1986 atmosphere at Mauna Loa, Hawaii (solid line):

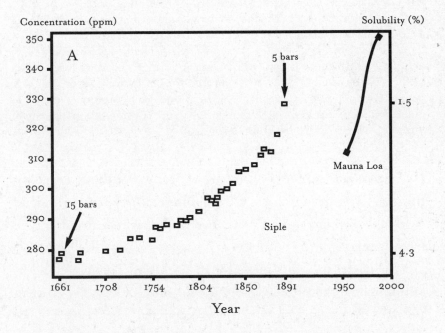

(A) Original Siple data, without assuming the air trapped within the ice is 83 years younger than the ice itself, suggests that CO2 levels in the 1890s were nearly as high as in the 1950s, undermining the CO2 emissions theory of global warming.

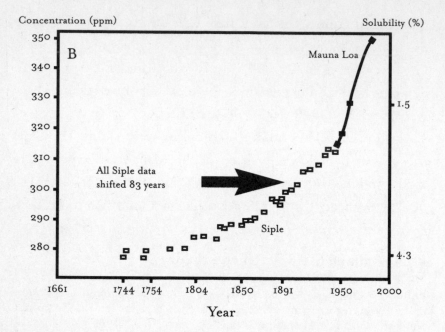

Concentration (ppm)   Solubility (%)

(B) The same data after arbitrary "correction" of the age of the air becomes consistent with the CO2 emissions theory of global warming. (Neftel et al., 1985; Friedli et al, 1986; and IPCC, 1990).

Adapted from Zbigniew Jaworowski, "Climate Change: Incorrect Information on Pre-Industrial C02," Statement prepared for the Hearing before the U.S. Senate Committee on Commerce, Science, and Transportation, March 19, 2004. Widely available on the net, one URL for Jaworowski's statement is: http://www.warwickhughes.com/icecore/.

But the rise in CO2 levels happens too early! For the industralization-as-CO2-driver and CO2-as-temperature-driver thesis to work, the Siple graph needs to move to the right, forward in time by 83 years.

So . . . the scientists moved it to the right! This was done by assuming that the average age of the air must be "exactly 83 years younger than the ice in which it was trapped. The 'corrected' ice data were then smoothly aligned with the Mauna Loa record (figure 1B) and reproduced in countless publications as the famous 'Siple curve.'"[10] It was not until years later, in 1993, that "glaciologists attempted to prove experimentally the 'age assumption'" but, says Dr. Jaworowski, "they failed."[11]

With the ascendancy of the IPCC and the creation of a huge funding pool for research advancing global warming, similar errors multiplied, he argues. "Improper manipulation of data and arbitrary rejection of readings that do not fit the pre-conceived idea on manmade global warming is common in many glaciological studies of greenhouse gases."[12]

Dr. Jaworowski understands glaciers and ice as few do: he has devoted much of his professional life to the study of the composition of the atmosphere as part of his work to understand the consequences of radioactive fallout from nuclear-weapons testing and nuclear-reactor accidents. After taking numerous ice samples over the course of a dozen field trips to glaciers in six continents, and studying how contaminants travel through ice over time, he came to realize how fraught with error ice-core samples are for purposes of reconstructing the history of the atmosphere. The Chernobyl accident, whose contaminants he studied in the 1990s in a Scandinavian glacier, was particularly illuminating.

"This ice contained extremely high radioactivity of cesium-137 from the Chernobyl fallout, more than a thousand times higher than that found in any glacier from nuclear-weapons fallout, and more than 100 times higher than found elsewhere from the Chernobyl fallout,"[13] he explained. "This unique contamination of glacier ice revealed how particulate contaminants migrated and also made sense of other discoveries I made during my other glacier expeditions. It convinced me that ice is not a closed system, suitable for an exact reconstruction of the composition of the past atmosphere."[14]

Because of the high importance of this realization, in 1994, Dr. Jaworowski, together with a team from the Norwegian Institute for Energy Technics, the University of Oslo, and the Norsk Polar Institute, proposed a research project on the reliability of trace-gas determinations in the polar ice. But the 30 prospective sponsors of the research, which included both public and private bodies, including several oil companies, (Elf Petroleum Norge A. S., Shell, Texaco, Fina, and Statoil) refused to fund Dr. Jaworowski's study. Even in 1993, they feared any association with research that might challenge the global warming orthodoxy, which was then much stronger in Europe than in the United States.

How Ice Cores Get CO2 Wrong

Adapted from the statement of Prof. Zbigniew Jaworowski written for the U.S. Senate Committee on Commerce, Science, and Transportation, March 19, 2004

Ice-core records of CO2 have been widely used as a proof that, due to man's activity, the current atmospheric level of CO2 is about 25% higher than in the pre-industrial period. These records became the basic input parameters in the models of the global carbon cycle and a cornerstone of the manmade climatic warming hypothesis. These records do not represent the atmospheric reality. . . .

Determinations of CO2 in polar ice cores are commonly used for estimations of the pre-industrial CO2 atmospheric levels. Perusal of these determinations convinced me that glaciological studies are not able to provide a reliable reconstruction of CO2 concentrations in the ancient atmosphere. This is because the ice cores do not fulfill the essential closed system criteria. One of them is a lack of liquid water in ice, which could dramatically change the chemical composition of the air bubbles trapped between the ice crystals. This criterion, is not met, as even the coldest Antarctic ice (down to -73°C) contains liquid water. . . . More than 20 physico-chemical processes, mostly related to the presence of liquid water, contribute to the

Why would the oil companies, in particular, refuse to fund scientific work by scientific institutions with impeccable reputations, especially when the future viability of the companies might rest on the knowledge that the agencies might discover? An oil company representative explained it at a meeting with Dr. Jaworowski's team. We have the record today because Dr. Jaworowski relayed his notes of the discussion shortly afterward to

alteration of the original chemical composition of the air inclusions in polar ice. . . .

One of these processes is formation of gas hydrates or clathrates. In the highly compressed deep ice, all air bubbles disappear, as under the influence of pressure the gases change into the solid clathrates, which are tiny crystals formed by interaction of gas with water molecules. Drilling decompresses cores excavated from deep ice and contaminates them with the drilling fluid filling the borehole. Decompression leads to dense horizontal cracking of cores, by a well-known sheeting process. After decompression of the ice cores, the solid clathrates decompose into a gas form, exploding in the process as if they were microscopic grenades. In the bubble-free ice the explosions form new gas cavities and new cracks. Through these cracks, and cracks formed by sheeting, a part of the gas escapes first into the drilling liquid that fills the borehole, and then at the surface to the atmospheric air. Particular gases, CO_2, O_2, and N_2 trapped in the deep cold ice start to form clathrates, and leave the air bubbles, at different pressures and depth. . . . This leads to depletion of CO_2 in the gas trapped in the ice sheets. This is why the records of CO_2 concentration in the gas inclusions from deep polar ice show the values lower than in the contemporary atmosphere, even for the epochs when the global surface temperature was higher than now. . . .[21]

a Polish friend. "I like the project, and we could easily cover all the expenses (about $2 million)," the oil company representative, explained. "We would gladly pay you the money provided that all other Norwegian oil and gas companies will also pay. Otherwise our competitors will report on us to the Norwegian government [headed by Mrs. Gro Brundtland, nicknamed the 'Environment Minister of the Globe'] and to Greenpeace. He feared there would be an outcry that his company supports 'an immoral project,' and this may induce a boycott of his company. This might cause much greater financial losses to the company than the costs of this project. His company consulted with a high level member of the government and got no green light, 'because that kind of a project is not to be accepted in Norway.'"[15]

The refusal from the 30 funding agencies, though a disappointment, did not come entirely as a surprise. Several years earlier, in a peer-reviewed article published by the Norwegian Polar Institute, Dr. Jaworowski had criticized the methods by which CO_2 levels were ascertained from ice cores and had explicitly cast doubt on the global warming hypothesis. The institute's director, while agreeing to publish his article, also warned Dr. Jaworowski that "this is not the way one gets research projects."[16] When Dr. Jaworowski's report was published, it soon sold out and was reprinted. The institute came under fire from critics of Dr. Jaworowski's report. Although none of the critics faulted Dr. Jaworowski's science, the institute did not extend his contract.[17]

To this day, the work that Dr. Jaworowski and his colleagues proposed has not been funded. But had it been funded, and had the results disproved the IPCC's ice-core assumptions, nothing much might have changed. After all, proxies apart from ice cores do exist; the IPCC rejects them. More tellingly, says Dr. Jaworowski, in most cases we "don't need to rely on proxies at all."[18] Scientists have been studying and measuring "CO_2 since the beginning of the 19th century, and they have left behind a record of tens of thousands of direct real-time measurements. These measurements tell a far different story about CO_2—they demonstrate, for example, that CO_2 concentrations in the atmosphere have fluctuated greatly and that several times in the past 200 years CO_2 concentrations have exceeded today's levels."[19] Nevertheless, he says, "the IPCC rejects these direct measurements, some taken by Nobel Prize winners. They prefer the view of CO_2 as seen through ice."[20]

CHAPTER EIGHT

Models and the Limits of Predictability

Hendrik Tennekes, Freeman Dyson,
Antonino Zichichi, David Bromwich

Antarctica is the world's most remote continent, its least explored and least understood, with a permanent ice sheet comprising more than 90% of the world's ice. Antarctica is also the primary heat sink in the global-climate system and "plays a central role in global-climate variability and change."[1]

Not until recently, with the advent of new technologies and improved scientific understanding, did human knowledge "allow the question of the global relevance of Antarctica to be explored in detail for the first time,"[2] says David Bromwich of the Byrd Polar Research Center at Ohio State University. In 1998, Dr. Bromwich began to take advantage of those new technologies in a major research project for the National Science Foundation in order to understand this frozen continent. His mission, in part, dealt with the science of global warming, which could not be settled until Antarctica gave up its mysteries.

A decade later, the unknowns surrounding the role of Antarctica

THE CV OF A DENIER

David Bromwich

David Bromwich is head of the Polar Meteorology Group of the Byrd Polar Research Center and professor in the Atmospheric Sciences Program at the Department of Geography of Ohio State University. He is president of the International Commission on Polar Meteorology, the chair of the Polar DAAC Advisory Group, a member of the Arctic Climate System Study Working Group on Reanalysis, and a past member of the National Academy of Sciences, Committee on Geophysical and Environmental Data. The author or coauthor of numerous papers, he received his Ph.D. in meteorology from the University of Wisconsin-Madison, in 1979.

in climate change continue to overwhelm the little that's known. As Dr. Bromwich reported in 2007 at the annual meeting of the American Association for the Advancement of Science at San Francisco, "It's hard to see a global warming signal from the mainland of Antarctica right now."[3] Antarctica's temperatures during the late 20th century did not climb as global-climate models—including those used by the IPCC—predicted. In fact, a 2006 study by Dr. Bromwich and others, published in the journal *Science*, found the accepted climate-change models to be wrong. According to the accepted models, snowfall in Antarctica should have been increasing. Instead, the study found that there has been no statistically significant increase in the snowfall trend over the past 50 years; snowfall patterns in Antarctica varied widely from year to year and decade to decade. "The best we can say right now is that the climate models are somewhat inconsistent with the evidence that we have for the last 50 years from continental Antarctica,"[4] he stated, adding that "we're looking for a small signal that represents the impact of human activity and it is hard to find it at the moment."[5]

★   ★   ★

In my research into the global warming controversy, no aspect of the doomsday case came in for so much criticism as the immensely complex and comprehensive computer models with which the IPCC and others claim to be able to predict climate change hundreds of years into the future. The case for manmade catastrophic global warming overwhelmingly depends on these models, yet even many of the supporters of the doomsday case seem to find the models—and especially the extreme claims made for their accuracy in the press and elsewhere—deeply embarrassing. Why?

A glimmer of the problem comes in a rather dramatic story about how one of Bromwich's own models helped save the life of Ronald Shemenski, a physician stationed at the U.S. South Pole Station in April 2001. Dr. Shemenski, who had developed a life-threatening pancreatic infection, needed to be airlifted in a season of high winds, extreme cold, and near 24-hour darkness, when planes rarely fly in the region. The rescue effort succeeded, thanks to a heroic aircrew, and also to Dr. Bromwich's model, which helped predict the best time for the perilous flight.

What's the difference between Bromwich's model and the global models used to predict global warming? Geographical focus, and time.

"The forecast model used to predict aircraft-landing conditions at the South Pole for the rescue was optimized specifically for Antarctic conditions,"[6] Dr. Bromwich explains. And the "model was only run for short periods, about two days at a time."[7]

Clearly, the prospects of success for a model that claims to predict the next hundred years are nothing close to the prospects of a model predicting the next 48 hours.

Perhaps even more important, the Antarctic is an extreme environment in which odd things happen. Models built to comprehend the entire Earth will not understand Antarctica very well. And since the fate of Antarctica is the single most important issue for global-climate change, getting Antarctica wrong means getting everything wrong.

"Global-climate models that are having some trouble at predicting the long-term behavior [over decades] of Antarctic near-surface temperatures are not optimized for the unique atmospheric conditions over Antarctica, probably the most pristine place on Earth,"[8] Bromwich explains. "The primary reason is connected with cloud formation. The global models treat the Antarctic clouds like those in mid-latitudes, whereas they are very different in reality."[9]

That global models fare poorly in remote parts of the world doesn't surprise him. "These are global models and shouldn't be expected to be equally exact for all locations,"[10] he explains. "Until the global models get the polar regions right, they won't get the global climate right either."[11]

In blunt layman's terms, the criticism of models that I encountered again and again comes down to this: global-climate models bite off more than they can chew. The result is not simply that the models occasionally make mistakes. That is to be expected even from quite good and useful models. Rather, as Hendrik Tennekes explains, the problem is that global-climate models reach a level of complexity so great that the predictions they issue can no longer be called scientific propositions.

*   *   *

Hendrik Tennekes, former director of research at the Royal Netherlands Meteorological Institute and later chairman of the august Scientific Advisory Committee of the European Center for Medium-Range Weather Forecasts, has touched an extraordinary number of lives in his own distinguished career, among academics and laymen alike. He is loved for his popular 1997 book, *The Simple Science of Flight From Insects to Jumbo Jets*, and for his scholarly 1972 work, *A First Course in Turbulence*, a classic that logs more than 2,000 citations on Google Scholar.

His most powerful influence has been on the practice of weather and climate modeling. As early as the 1980s, Tennekes was challenging the models that climate scientists were constructing, saying models could never replicate the complexity of the real world.

Tennekes issued his first great warning on the subject in a landmark 1986 speech in Riga, Denmark, "No Forecast Is Complete Without A Forecast of Forecast Skill."[12] He argued that meteorology was being put in a position "at odds with the mainstream of the scientific enterprise of the last 300 years.... The goal of science is prediction, but we stand in front of the limits of predictability."[13] The standard for an acceptable scientific hypothesis

THE CV OF A DENIER
Hendrik Tennekes

Hendrik Tennekes is a member of the Royal Netherlands Academy of Arts and Sciences. He studied aeronautical design at TU Delft in the Netherlands and was a professor of aerospace engineering at Pennsylvania State University, a professor of meteorology at Free University, Amsterdam, and director of research at the Royal Netherlands Meteorological Institute. Born in 1936, he still lectures at the University of Nijmegen, the Netherlands.

has long been that the hypothesis makes predictions that can be subsequently tested. Meteorology, argued Tennekes, would soon hit a brick wall in this respect. He argued that modern theory "unequivocally predicts that no amount of improvement in the quality of the observation network or in the power of computers will improve the average useful forecast range by more than a few days."[14] Since Tennekes's speech, a host of weather scientists have sought to extend the bounds of modeling. They have seen success, but only on the scale Tennekes predicted.

In an attempt to save and extend the weather-modeling enterprise in the face of Tennekes's warnings, scientists developed the now-common disciplines of "ensemble forecasting" and "multi-model forecasting." Tennekes has not been much impressed, calling ensemble forecasting "a poor man's version of producing a guess at the probability density function of a deterministic forecast."[15] The new approaches to modeling soon spread beyond weather forecasting to climate research too. "But fundamental questions concerning the prediction horizon are being avoided like the plague."[16]

The problem, Tennekes eventually realized through the work of the great 20th-century philosopher of science Karl Popper, was a profound one. Popper had struggled against an ideology he called scientific determinism, which states that if we knew everything there was to know about the current physical state of the world, we could perfectly predict all its future states. Fascinatingly to Tennekes, Popper used clouds as a way of explaining the problems of determinism.

"Popper . . . introduced clouds to represent physical systems 'which are highly irregular, disorderly, and more or less unpredictable,' and clocks to represent systems 'which are regular, orderly, and highly predictable in their behavior.' Popper's interest in predictability was fueled by his opposition to the views held by the proponents of determinism. The central thesis of determinism is the 'staggering proposition' that all clouds are clocks; in this view the distinction between clocks and clouds is not based on their intrinsic nature, but on our lack of knowledge. If only we knew as much about clouds as we do about clocks, clouds would be just as predictable as clocks. Or, in meteorological terms, a perfect model of the atmosphere, initialized with perfect data from an observation network of infinite resolution, and run on an infinitely powerful computer, should in principle produce a perfect forecast with an unlimited range of validity."[17]

Popper saw determinism as not only false but dangerous, leading to undisciplined, arrogant, and worst of all unfalsifiable predictions masquerading as science. Popper's famous principle that a scientific proposition must be "falsifiable" requires that every scientific claim include at least implicitly a clear notion of what evidence would be sufficient to prove that the proposition is untrue. If I claim there is an elephant under my bed, I have actually made a scientific claim (hopefully a false one) because it can be tested—by looking under my bed. But if I claim there is an invisible fairy elephant under my bed, I have not made a scientific claim because it cannot be disproved.

The danger of determinism as Popper saw it, and as Tennekes would come to understand it in the context of climate studies, is that it would lead to unfalsifiable theories that looked like science but weren't. The principle that perfect knowledge would lead to perfect predictions would create an escape route for scientists who, like climate modelers, sought to explain extremely complex physical systems by building ever-more complicated models to approximate the systems. Of course the information plugged into the models could never be perfect. Thus, when the model made a prediction that turned out to be false, the model itself could always be defended on the grounds that the data fed into the model had been faulty or insufficient. The model itself—which is in effect the proposition or theory—would not be falsifiable because the modeler could always claim that with perfect information it would have worked. As Tennekes explains:

"[Popper is saying] we have to agree in advance on the accuracy we demand. We should not allow ourselves an escape route if things go wrong. . . .

". . . Whenever they fail in their predictions, scientists tend to blame the poor accuracy of the observations, the lack of computer power, and the inadequate parameterization in their numerical models, rather than their own lack of skill in computing the accuracy that can be obtained with present resources. Sloppy reasoning of this kind is responsible for much of the thoughtless expansion and escalation numerical modelers in all branches of science indulge in.

". . . To put it bluntly, a calculation that does not include a calculation of its predictive skill is not a legitimate scientific product. . . . [As Popper puts it] 'we must be able to account *in advance* for any failure to predict an event with the desired degree of precision, by pointing out that our initial conditions are not precise enough, and by stating how precise they would have to be for the particular prediction task at hand.'

"Popper [points out] that accountability is concerned not only with the accuracy of the initial conditions but also with the limitations inherent in the models we employ to describe the evolution of the physical system: 'The method of science depends on our attempts to describe the world with simple theories. Theories that are complex may become untestable, even if they happen to be true. Science may be described as the art of systematic oversimplification, the art of discerning what we may with advantage omit.'"[18]

Climate is so complex and so many simplifications are required that many models quickly become useless as predictive devices. Tennekes states, for instance, that we cannot possibly predict future climate changes unless we can predict changes in precipitation. But this requires an understanding of a number of complex interactive forces far beyond anything we have today and possibly beyond our theoretic ability to achieve.

Why is it so difficult to make precipitation forecasts 50 years into the future? "We do not know, and for the time being cannot know anything about changing patterns of clouds, storms, and rain. Holland's national weather service KNMI circumvented this impasse last year by issuing climate-change scenarios with and without changes in the position of the North Atlantic storm track. It did not occur to the KNMI spokesmen that they should have been forthright about their lack of knowledge. They should have said: we know nothing of possible changes in the storm track, so we cannot say anything about precipitation. But it is entirely consistent with the IPCC tradition to weasel around such issues. One of my contacts at KNMI recently explained to me that their choice was based on the increasing agreement between simulations run with different GCMs [General Circulation Models]."[19]

Translation: in the sort of climatology that the IPCC employs, the results of a model that necessarily ignores fundamental issues are checked against the results of other models that necessarily ignore the same issues. If several such models, all incomplete in the same way, are seen "increasingly" to agree, then we are supposed to accept the results.

This explains the "IPCC's preoccupation with CO2": "Sophisticated climate models have been running for twenty years now. It has become evident that these models cannot be made to agree on anything except a possible relation between greenhouse gases and a slight increase in globally averaged temperatures. . . ."[20] Naturally, modelers focus on the few things they know rather than the multitude they do not. But in doing so they have become the pawns of politicians. "We should stop our support for the preoccupation with greenhouse gases our politicians indulge in. Global energy policy is their business, not ours. We should not allow politicians to use fake doomsday projections as a cover-up for their real intentions."[21]

Summing up his doubts, Tennekes explains that the "blind adherence to the harebrained idea that climate models can generate 'realistic' simulations of climate is the principal reason why I remain a climate skeptic."[22] "There exists no sound theoretical framework for climate predictability studies."[23]

But if we cannot predict climate change, what are we to do? After all, global warming could be real. Should we simply abandon ourselves to fate?

Hardly, says Tennekes. If we cannot predict events, however, then prediction cannot be our weapon of choice in dealing with them. If we cannot predict change, we must focus on adapting to it when it comes. If we cannot predict the future, we must be ready to cope with the unpredictable. As Tennekes put it in one of his most quotably oracular moments:

"The constraints imposed by the planetary ecosystem require continuous adjustment and permanent adaptation. Predictive skills are of secondary importance."[24]

Is it realistic, or even responsible, to say, oh well, we'll deal with the future when it gets here? Actually, we do it all the time when confronting complexity. Modern military doctrine, for instance, emphasizes communication and rapid adaptation to shifting circumstances, pushing information and decision making down to the lowest possible levels so that units and even individual soldiers can deal with the unexpected as it happens. Battlefield events have become too fast moving and complex to model in advance.

A more precise analogy comes from sports. Consider football

in contrast to soccer or basketball. In football the actions of each team member for any given play are elaborately planned, at least for the offense. This is feasible only because the average play continues for only a few seconds. When a play lasts much longer, it is almost always because something chaotic has occurred, like an interception. When that happens both teams not only find themselves off plan but actually switching sides from offense to defense and vice versa. In soccer and basketball, play is continuous, keeping "play calling" to a minimum. The longer time horizon increases complexity and makes planning—or forecasting—futile.

Which of course is not to say that modern generals, or soccer coaches, have nothing to do but await events. On the contrary, the effective modern soldier is far more elaborately trained and prepared than his 18th-century counterpart. He must be better trained precisely because his job is to respond to uncertain developments. Tennekes reminds us that the most important thing to know about climate is what we do not know—and for that very reason, we must be ready to respond to the unexpected.

His reward for these efforts? He has become an object lesson in the politically acceptable limits of scientific inquiry. Because his critiques of climate science ran afoul of the orthodoxy required by the Royal Netherlands Meteorological Institute, he was forced to leave. Lesser scientists, seeing that even a man of Tennekes's reputation was not free to voice dissent, learned their lesson. Those who harbor doubts about climate science do better to bite their tongues and keep their heads down.

Fighting Climate "Fluff" with Freeman Dyson

Physicist Freeman Dyson knows from long experience that models packed with numerous "fudge factors" are worthless.

As a mathematician and physicist, Dyson is known for the unification of three versions of quantum electrodynamics, for his work on the Orion Project, which proposed space flight using nuclear pulse propulsion, and for developing the TRIGA, a small, inherently safe nuclear reactor used by hospitals and universities worldwide for the production of isotopes.

As a theoretician, he is known for the Dyson sphere (an inspiration for science fiction such as *Star Trek*, as well as scientific works), the Dyson transform (which led to the discovery that every even integer is a sum of at most six primes), and the Dyson tree (a genetically engineered plant capable of growing on a comet).

As an activist and visionary, he is known for his concern for global poverty, for his promotion of international cooperation, and for his work in furtherance of nuclear disarmament. He is a member of the board of sponsors of the Bulletin of Atomic Scientists and the subject of numerous writings by environmental pioneers, such as Stewart Brand.

But these days this Renaissance man is known as a scientific heretic, chiefly for disagreeing with, as he puts it, "all the fluff about global warming."[25]

As with Henk Tennekes, Dyson's skepticism about the science is driven by his skepticism about the models. "I have studied their climate models and know what they can do,"[26] Professor Dyson says. "The models solve the equations of fluid dynamics and do a very good job of describing the fluid motions of the atmosphere and the oceans. They do a very poor job of describing the clouds, the dust, the chemistry, and the biology of fields, farms, and forests. They do not begin to describe the real world that we live in."[27]

Professor Dyson argues that the many components of climate models reflect a poor understanding of first principles and thus can-

THE CV OF A DENIER
Freeman Dyson

Freeman Dyson, a graduate of Cambridge University in 1945 with a B.A. degree in mathematics, has been for most of his life a professor of physics at the Institute for Advanced Study in Princeton. He is a fellow of the American Physical Society, a member of the U.S. National Academy of Sciences, and a fellow of the Royal Society of London. In 2000, he was awarded the Templeton Prize for Progress in Religion.

not capture the effects of change in complex interactive systems. "They are full of fudge factors that are fitted to the existing climate, so the models more or less agree with the observed data. But there is no reason to believe that the same fudge factors would give the right behavior in a world with different chemistry, for example, in a world with increased CO_2 in the atmosphere,"[28] he states.

"Typical examples of fudge factors occur in the treatment of clouds. Each cell of the atmosphere in the model is characterized by a set of numbers which specify the temperature, pressure, density, humidity, wind velocity, cloudiness, etc., in that cell. Since the cell is much larger than a typical cloud, the 'cloudiness' number is only a rough measure of the fraction of the cell that is occupied by clouds. An empirical formula then gives the rate of precipitation in the form of rain or snow for a cell with a given humidity and given cloudiness. The empirical formula contains several coefficients that are fitted to the observations to make the model agree with the existing climate. These coefficients are what I call 'fudge-factors.' They are not based on a detailed understanding of clouds and rainfall but only on fitting a formula to observations.

If now the model is run with enhanced CO2, there is no reason to believe that the same fudge factors will still give the right answers. There are many other fudge factors concerned with processes such as snow melting and vegetation-growth that cannot be modeled in detail."[29]

Unlike many scientists today, who seek the comfort of consensus as opposed to thinking for themselves, Prof. Dyson has always been willing to be in the minority and to convey uncomfortable truths. He tells the story of his stint as an analyst during the Second World War in the UK's Bomber Command, when he proposed ripping out two gun turrets from RAF Lancaster bombers. Without the turrets, they could fly 50 miles per hour faster, be much more maneuverable, and cut the UK's catastrophic losses to German fighters. Those at the top preferred to

Freeman Dyson:

Science and politics require different standards for belief

From "The Science and Politics of Climate" by Freeman J. Dyson, APS [American Physical Society] News, May 1999

In the nineteen-sixties the fluid dynamicist Syukuro Manabe was running global-climate models on the super-computer at the Geophysical Fluid Dynamics Laboratory in Princeton. Manabe began very early (before it became fashionable) to run models of climate with variable amounts of carbon dioxide in the atmosphere. He ran models with carbon dioxide at two and four times the present abundance, and saw in the computer output, the rise in average ground temperature that is now called Global Warming. He told everybody not to believe the numbers. . . .

delude themselves: "To push the idea of ripping out gun turrets, against the official mythology of the gallant gunner defending his crew mates, was not the kind of suggestion the commander in chief liked to hear."[30]

Today's official mythology involves global warming, in a societal mobilization of another kind, but the duty remains the same. "Here I am opposing the holy brotherhood of climate model experts and the crowd of deluded citizens that believe the numbers predicted by their models."[32] A heretic he remains, and as history has shown, much more often right than not.

"I'm not saying the warming doesn't cause problems. Obviously it does. Obviously we should be trying to understand it. I'm saying that the problems are being grossly exaggerated. They take away money and attention from other problems that are much

Over and over again he said that his purpose when he ran computer models was not to predict climate but to understand it. But nobody listened. Everyone thought he was predicting climate, everyone believed his numbers . . . the politicians in Washington believed. They wanted numbers, he gave them numbers, so they naturally believed the numbers.

It was not unreasonable for politicians to believe Manabe's numbers. Politics and science are two very different games. In science, you are not supposed to believe the numbers until you have examined the evidence carefully. If the evidence is dubious, a good scientist will suspend judgment. In politics, you are supposed to make decisions. Politicians are accustomed to making decisions based on shaky evidence. They have to vote yes or no, and they generally do not have the luxury of suspending judgment. Manabe's numbers were clear and simple. They said if the carbon dioxide goes up, the planet will get warmer. So it was reasonable for politicians to believe them. Belief for a politician is not the same thing as belief for a scientist.[31]

more urgent and more important—poverty, infectious diseases, public education, and public health. Not to mention the preservation of living creatures on land and in the oceans."[33]

Faith and Reason in Rome

Dr. Antonino Zichichi is Italy's most renowned scientist. He has published more than 800 scientific papers, opening new avenues in Subnuclear Physics at High Energies. He is credited with the discovery of nuclear antimatter; the discovery of the "time-like" electromagnetic structure of the proton; the conjecture of the existence of a third lepton and the invention of new technologies which led to the discovery of the third family in the structure of the fundamental particles; the discovery of the effective energy in the forces which act between quarks and gluons; and the proof that, despite its complex structure, it is impossible to break the proton.

Zichichi is also an outspoken critic of the IPCC. He delivered

Freeman Dyson on the Four Reservoirs of CO2

From "The Science and Politics of Climate" by Freeman J. Dyson, APS [American Physical Society] News, May 1999

The biosphere of the Earth contains four reservoirs of carbon: the atmosphere, the ocean, the vegetation, and the soil. All four reservoirs are of comparable size, so that the problem of climate is inescapably mixed up with the problems of vegetation and soil. The intertwining between the four reservoirs is so strong that it makes no sense to consider the atmosphere and ocean alone. Computer models of atmosphere and ocean, even if they can be made reliable, give at best a partial view of the problem. The large effects of vegetation one of his most damning critiques to date at a 2007 Vatican conference on "Climate Change and Development," designed to "search for solutions to the phenomenon of global warming."[34] Eighty scientists, politicians, theologians, and bishops were in attendance. During the two-day event, tensions often ran high. No one left the seminar thinking that the science of global warming was settled. To the dismay of those hoping that the high-level group would inspire a Church-led climate-change crusade, Cardinal Renato Martino, who chaired the event, in closing urged caution in taking any position on global warming.

Perhaps the chief reason for this restraint was Dr. Zichichi, who argued "that models used by the Intergovernmental Panel on Climate Change (IPCC) are incoherent and invalid from a scientific point of view. . . . On the basis of actual scientific fact 'it is not possible to exclude the idea that climate changes can be due to natural causes,' and that it is plausible that 'man is not to blame.'"[36]

and soil cannot be computed but must be observed and measured.

The way the problem is customarily presented to the public is seriously misleading. The public is led to believe that the carbon dioxide problem has a single cause and a single consequence. The single cause is fossil-fuel burning; the single consequence is global warming. In reality there are multiple causes and multiple consequences. The atmospheric carbon dioxide that drives global warming is only the tail of the dog. The dog that wags the tail is the global ecology: forests, farms, and swamps, as well as power stations, factories, and automobiles. And the increase of carbon dioxide in the atmosphere has other consequences that may be at least as important as global warming—increasing crop yields and growth of forests, for example. To handle the problem intelligently, we need to understand all the causes and all the consequences. . . .[35]

Dr. Zichichi believes that solar activities are responsible for most of the global warming that Earth has experienced, with manmade causes accounting for only a small portion of warming. His conclusions have gravitas: this man is the president of the World Federation of Scientists, past president of the NATO Science Committee for Disarmament Technology, past president of the European Physical Society, and past president of the Italian National Institute for Nuclear and Subnuclear Physics.

Like many physicists, he is appreciative of the analytic power of quantitative models, but apprehensive of their potential for becoming self-reflexive and unhinged from reality. That awareness has made him a penetrating critic of the IPCC for its overreliance on underproven models.

Zichichi observes that even in particle physics, with all its intrinsic rigor and capacity for controlled experimentation, "we have a lot of problems in making predictions."[37] Extrapolating those same challenges to meteorology and climatology, which are of necessity, "far less rigorous than the study of the Fundamental Interactions,"[38] suggests that "predictions in meteorology and climate change must necessarily be taken with great caution."[39] "We should avoid giving to the public the message that science has mastered all meteorology and climate-change problems of the past, present, and future,"[40] he said, noting that unfortunately the public has concluded the same.

Climate scientists have achieved a great deal, he agreed. But "[t]he mathematics involved [in climate modeling] is a system of strongly coupled, non-linear differential equations, where the solution can only be arrived at by a series of numerical approximations. In these approximations you need to introduce a number of free parameters. [John] von Neumann"—the great 20th-century mathematician—"was always warning his young collaborators about the use of these free parameters by saying: If you allow me

THE CV OF A DENIER
Antonino Zichichi

Antonino Zichichi, professor emeritus of Advanced Physics at the University of Bologna, has published over 800 scientific papers and ten books, some of which have opened new avenues in subnuclear physics. He founded and directs the Ettore Majorana Foundation and Center for Scientific Culture, an organization dedicated to voluntary scientific service, the elimination of secret laboratories, and scientific freedom. Zichichi is president of the Enrico Fermi Center, Rome, and president of the World Federation of Scientists, with 10,000 scientists from 115 countries. He has served as president of the INFN (Italian National Institute for Nuclear and Subnuclear Physics); president of the EPS (European Physical Society); and president of the NATO Science Committee for Disarmament Technology (nuclear, chemical, bacteriological, and conventional). Zichichi has received awards and honorary degrees from universities and academic institutions in Italy and in many other countries, including Argentina, China, Georgia, Germany, Poland, and the United States.

four free parameters I can build a mathematical model that describes exactly everything that an elephant can do. If you allow me a fifth free parameter, the model I build will forecast that the elephant will fly."[41]

The difficulty with the mathematical models for climate change, he explains, is that they "have a lot more than five free parameters. There is a minimum of two for each volcano. And then those necessary to describe the dynamic properties of the air strip that surrounds this sun's satellite, with all the interactions between atmosphere, ocean, winds, maritime currents, and greenhouse gases. There are also the free parameters for the particles of dust, soot, and other substances being constantly injected in the atmosphere, without the possibility of an accurate check of their characteristics, whether in terms of quantity or quality. These 'dust' particles play an important role in the thermodynamics of the atmosphere."[42]

Even to build a contemporary model of the interactions of all these factors would be a staggering effort. But the problem is worse than that. For a model to predict global temperature resulting from contemporary human activities, modelers would need accurate models of how these factors operated in the past. "[W]e would need to know the variation with time of these particles injected in the atmosphere, including aerosols, for which it would be important to know what was happening in the past, before we had specific measuring instrumentation and therefore know little or nothing of."[43]

Furthermore, variations in climate involve not just the atmosphere but the physics and chemistry of the earth and the seas as well. "[H]istory has taught us that phenomena of strong variation occurred, which resulted in the transformation of magnificent expanses of green land—such as for Greenland—into vast expanses of ice, and luxuriant extensions of vegetal life into deserts, such as the Sahara today. If we ever come up with a mathematical structure capable of describing the past of the solid and liquid surfaces of earth, and only then, it will be possible to confirm what is being advocated today by the 2,500 scientists of the Intergovernmental Panel for Climatic Change (IPCC)."[44]

For these purposes, he argues, "The present mathematical models are far from being satisfactory. The public at large wishes to know if it is true that human activities are creating a huge perturbation of the climate characteristics of our globe."[45] The UN IPCC, he lamented, "has led the public to believe . . . that science has understood all about climate. . . . But it's not this way."[46]

If the solutions that global warming doomsayers are demanding were free or without grave consequence for the world's poor, then perhaps "better safe than sorry" would be an acceptable policy. But the Vatican's Pontifical Council for Justice and Peace, which sponsored the conference, has concerned itself with global warming precisely because both the alleged problem and the proposed solutions could have grave consequences for the poor—especially in underdeveloped countries that have been the Church's special concern. For that reason the conference was specifically asked to consider whether "[g]lobal warming may bring about not only the imposition of drastic corrective means to protect the natural environment, but also a grave threat that destabilizes the world."[47]

Dr. Zichichi is one of the co-founders of the World Federation of Scientists, a group with more than 10,000 members in 110 countries whose purpose is to focus science on "planetary emergencies" from climate change to terrorism, from missile proliferation to the health and safety of mothers and children, with special emphasis on underdeveloped nations that lack either political influence to affect change, or the economic strength to cope with these emergencies. Thus, Dr. Zichichi is deeply aware that global warming is only one alleged calamity facing the world's poor, and that every dollar and even more important *every hour* of scientific attention diverted from real crises to a possibly phony one has real and tragic costs. In the end, his concern with the defects of global warming models is not primarily with their mathematical and empirical shortcomings but with the immense human costs those shortcomings may impose. "We're talking here of mathematical models whose results have consequences costing billions of dollars and involve the responsibility of all the governments in the world. It is necessary to bring these basic themes back to the scientific laboratories where they belong, taking them away from the hands of those who use them to satisfy ambitions that have nothing to do with scientific truth."[48]

$\star$ $\star$ $\star$

Perhaps the most powerful, recent articulation of the limitations of climate models has come from a rather stunning source, Dr. Kevin Trenberth. Dr. Trenberth is the IPCC official whose hurricane hype prompted Christopher Landsea's resignation from the IPCC (see chapter three) and who has been described by Prof. Bob Carter as "one of the advisory high priests of the UN's Intergovernmental Panel on Climate Change."[49] Dr. Trenberth recently argued on "Climate Feedback: The Climate Change Blog," published by the journal *Nature*, that IPCC models do not and cannot predict future climate. That, he says, is not even their purpose. "In fact there are no predictions by IPCC at all. And there never have been. The IPCC instead proffers 'what if' projections of future climate that correspond to certain emissions scenarios."[50]

The models are not capable of predicting future climate. By design, they omit crucial information, as Trenberth explains:

"None of the models used by IPCC are initialized to the observed state, and none of the climate states in the models correspond even remotely to the current observed climate. In particular, the state of the oceans, sea ice, and soil moisture has no relationship to the observed state at any recent time in any of the IPCC models. There is neither an El Niño sequence nor any Pacific Decadal Oscillation that replicates the recent past; yet these are critical modes of variability that affect Pacific rim countries and beyond. The Atlantic Multidecadal Oscillation, that may depend on the thermohaline circulation and thus ocean currents in the Atlantic, is not set up to match today's state, yet it is a critical component of the Atlantic hurricanes, and it undoubtedly affects forecasts for the next decade from Brazil to Europe. Moreover, the starting climate state in several of the models may depart significantly from the real climate owing to model errors. I postulate that regional climate change is impossible to deal with properly unless the models are initialized."[51]

Since he made this statement, Trenberth and others have charged that it has been quoted selectively and distorted by some global warming critics to make it appear that Trenberth thinks the limitations of the models cast doubt on human-caused global warming. He clearly does not think this. In the very same statement, he makes it perfectly clear that he accepts the IPCC's opinion that the "warming of the climate system is unequivocal"[52] and that he believes "the science is convincing that humans are the cause."[53] Moreover, he clearly believes that the models accepted by the IPCC have made serious contributions to the understanding of climate change, something many skeptics might say as well. Because the statement has become so controversial I urge readers to read it in full at: http://blogs.nature.com/climatefeedback/categories/topics/climate_science/climate_variability/.

In the end, however, Trenberth essentially supports what critics of climate models say. His list of the models' defects resonates with the complaints of critics like Bromwich, Tennekes, Dyson, Zichichi, and others. And though he says that the evidence that humans are the cause of global warming is "convincing," his argument for human responsibility does not depend on the complex global-climate models whose weaknesses he elucidates.

In the mind of the public, however, the models have been huge. For it is largely from the models that the great and terrifying doomsday scenarios have come. The public is never told that these horror stories are merely "stories," or speculative "what if " scenarios. Just the opposite. It is the models and the stories they tell that have convinced much of the public that the consequences, if global warming does come true, will be so grave that we really should be better safe than sorry Meanwhile, as Zichichi and others point out, most people have little idea what the cost of such "safety" might be—especially to the poor of the world.

This is what the models have done: turned the future into a horror story. And for that reason the news that Kevin Trenberth— and a good many others in the doomsday camp—hold a rather dim view of the models' ability to predict that the future is a big story indeed.

CHAPTER NINE

In the Land of the Midnight Sun

Eigil Friis-Christensen, Henrik Svensmark

Nir Shaviv, whom we met in chapter six, is eloquent on the question of why so many scientists, himself included, so quickly accepted the hypothesis that CO_2 emissions were the dominant cause of global warming. It really came down to a very simple process of elimination.

We know there is a greenhouse effect—without it life on Earth would be impossible—and that changes to CO_2 affect the amount of energy in Earth's atmosphere.

We know that industrial society is pumping a good deal of CO_2 into the air.

And—perhaps the most important point—we know, or knew, very little else about why Earth's temperature changes over decades, centuries, or eons.

Nor was Shaviv referring to only his own personal ignorance, the fact that as an astrophysicist climate was not his specialty. He was giving a perfectly accurate summary of the state of climate knowledge as of the time the global warming movement got off the ground—in the early 1990s.

Considered in isolation, we understand the greenhouse effect pretty well. We can do controlled experiments in the lab that reproduce aspects of it. And, despite the huge ongoing arguments about the data, we can do some rough reconstruction of the history of CO_2 concentrations in the atmosphere and global temperatures going back hundreds of millions of years and look for correlations between CO_2 levels and temperature levels.

What we know much less about is how changes in CO_2 affect the climate overall. And, leaving CO_2 aside for the moment, we know vanishingly little about how the Sun—our main source of energy driving the climate—affects climate change.

Most laymen would probably find this surprising. I know I did. Our directly observed records of the Sun's behavior are scanty. Our "instrumental" record of factors like total solar irradiance or changes in the Sun's magnetic field go back only a few decades. We can do reconstructions based on proxies and indirect evidence. But this is even harder to do for the Sun than for CO_2 levels or temperature.

In the late 1980s and early 1990s, astrophysicists and other space scientists began assembling the fractional and frustrating instrumental record on the Sun and extending it—where they could—back into pre-instrumental times. Using the emerging record, they began looking for intriguing correlations that might hint at actual physical explanations of solar behavior and its implications for climate.

Naturally, this early research was not especially focused on the global warming debate, which was just getting started. Yet some of the solar scientists did hope that the rising interest in global warming might mean that the UN and the IPCC would help push for solar-climate research to help unlock the mysteries of climate change.

What really happened was just the reverse. The solar scientists found themselves being identified as enemies as far as the growing global warming establishment was concerned. Rather than funding and encouraging their research, officials of the IPCC and other leading forces in the global warming movement began denouncing the solar scientists as irresponsible and worse. Institutes funding solar-climate research faced furious criticism; funding for crucial experiments was delayed or cut off. Worst of all was a campaign to smear the solar researchers personally, even charging them with deliberately distorting data in the service of special interests.

Of course there have been legitimate scientific critiques of the solar theorists' research and analysis—most of them from the solar scientists themselves. And neither in my columns nor in this book have I ever made any attempts to adjudicate scientific disagreement. But even without "judging the science," a layman can distinguish fair from foul play, especially when the foul has been so blatant. Perhaps the worst tactic used against the solar theorists has been to portray honest revision as trickery and flip-flopping.

For instance, rather dramatic graphs have been published demonstrating that certain apparently powerful correlations between solar activity and climate, which prompted some of the early solar hypotheses, had suddenly stopped working. Where were these graphs published? In the papers of the solar theorists themselves! Yet in the propaganda campaign against the solar theorists, this steady progress of understanding, this almost classic picture of how science is supposed to proceed has been portrayed as essentially dishonest. A "Pattern of Strange Errors,"[1] as one critic put it.

Vicious as the campaign against all deniers has been, nothing can compare to the vitriol directed at the solar theorists. For as

Shaviv implies, a coherent theory of solar influence on current climate change is potentially far more devastating to Al Gore's claim that the science is settled than any one criticism of the manmade CO2 hypothesis itself.

The media assault on denial rarely mentions the solar scientists by name. The infamous *Newsweek* story "The Truth About Denial" left them in the background, preferring to focus on "the denial machine," the network of lobbyists and (typically) conservative journalists who cite (sometimes selectively and improperly) the work of the solar theorists so as to attack Gore and the CO2 machine.[2] It is only by avoiding the scientists and focusing on polemicists that someone like Scott Pelley of *60 Minutes* can say something like, "If I do an interview with Elie Wiesel, am I required as a journalist to find a Holocaust denier?"[3] Only by pretending that serious theorists don't exist can Al Gore get away with quips about how global warming deniers "get together on a Saturday night and party with"[4] people who believe that the Earth is flat and that the moon landing was staged on a movie lot.

So who are these flat-earthers? And where do they work?

A complete list of every qualified scientist today actively engaged in research relevant to the role of the Sun or astrophysics in climate change would easily run to the hundreds and perhaps the thousands. A list of institutions at which their work is done would include most of the world's important centers of astrophysical research.

In pursuing this story, I have encountered or been referred to dozens of scientists who have done truly significant work in the field. But just as a sampling, before we get into the details of the work, here is a list of just ten of the more prominent scholars in the field and their institutions. See if the first words that come to your mind are "crackpots," "out of the scientific mainstream," "naïve," or "irresponsible."

- Dr. Habibullo Abdussamatov, head of the space research laboratory of the Pulkovo Observatory and of the International Space Station's Astrometry project at the Russian Academy of Science
- Dr. Eigil Friis-Christensen, director of the Danish National Space Center
- Freeman Dyson, past professor of physics at the Institute for Advanced Study in Princeton and a member of the U.S. National Academy of Sciences
- Dr. Rhodes Fairbridge of Columbia University and editor of the *Encyclopedia of Earth Sciences*
- Dr. Michael Griffin, administrator (CEO), National Aeronautical and Space Association (NASA)
- Dr. Jasper Kirkby, particle physicist and leader of the CLOUD research project at CERN—the European Organization for Nuclear Research
- Dr. Sami Solanki, director and scientific member at the Max Planck Institute for Solar System Research in Germany
- Dr. Henrik Svensmark, director of the Center for Sun-Climate Research at the Danish National Space Center
- Lev Zeleny, director of the Institute of Space Research at the Russian Academy of Sciences
- Antonino Zichichi, professor emeritus of advanced physics at the University of Bologna and president, World Federation of Scientists

That's not a bad list. Here's another way to slice the data, by looking at a list of the research institutes involved with just one of the experimental programs underway to explore the physical mechanisms linking the Sun to climate change—the CLOUD experiment, which began in 2007 and is expected to run through 2012.[5]

- University of Bergen, Norway
- California Institute of Technology, United States
- CERN, Switzerland

- Danish National Space Center, Denmark
- Finnish Meteorological Institute, Finland
- University of Helsinki, Finland
- University of Kuopio, Finland
- Lebedev Physical Institute, Russia
- University of Leeds, United Kingdom
- Leibniz Institute for Tropospheric Research, Germany
- University of Mainz and Max Planck Institute for Chemistry, Germany
- Max Planck Institute for Nuclear Physics, Germany
- Paul Scherrer Institute, Switzerland
- University of Reading, United Kingdom
- Rutherford Appleton Laboratory, United Kingdom
- Tampere University of Technology, Finland
- University of Vienna, Austria

So much for the politics. But is there a serious alternative theory? The indisputable answer is *yes*. Many if not most of the world's most prestigious research institutions are dedicating resources to investigating the role that planetary and cosmic forces may play in Earth's climate. Whether or not these theories emerge triumphant we cannot yet know. I do know that this story deserves to be told.

We start with Dr. Eigil Friis-Christensen, perhaps the most thoroughly reviled and condemned denier of all. Who is he? To begin with, he is the chairman of the Danish Space Consortium. He heads a European Space Agency mission advisory group. He is vice president of the International Association of Geomagnetism and Aeronomy. For his scientific rigor, Dr. Friis-Christensen has won a citation from the *Journal of Geophysical Research of the American Geophysical Union* for "Excellence in Refereeing." He is sought after by the world's space agencies, which have elevated him to the top ranks of his profession. Many of the world's most

THE CV OF A DENIER
Dr. Eigil Friis-Christensen

Eigil Friis-Christensen is director of the Danish National Space Center. In the European Space Agency, ESA, he was a member of the Solar System Working Group from 1994 to 1998 and has been a member of the ESA Science Program Committee since 1998. From 2003 to 2007, he was a member of the Earth Science Advisory Committee (ESAC), and he is currently the chairman of the Mission Advisory Group (MAG) for Swarm—"The Earth's Magnetic Field and Environment Explorers" in the Earth Observation Program of ESA. He has been a member of the Space Research Advisory Committee of the Swedish National Space Board from 1998 to 2006. From 2003 to 2006, he was member of NASA's "Living With a Star," Management Operations Working Group. He is author or coauthor of some 100 peer-reviewed articles and has presented more than 50 invited papers at international conferences. He holds a Magisterkonferens (Ph.D.) in geophysics from the University of Copenhagen.

prestigious space-related research institutions—the European Organization for Nuclear Research in Geneva, the Max Planck Institute for Solar System Research in Germany, and the Pulkovo Astronomical Observatory in Russia among them—are now building on the solar theories of climate change that Dr. Friis-Christensen set in train.

Dr. Friis-Christensen's interest in climate change predates the Kyoto Treaty of 1995, it predates the Rio Conference in 1992 that led to Kyoto, it even predates the first report in 1990 of the IPCC.

"My interest dates back to an extreme solar storm that occurred in August 1972,"[6] he explains. "I was in Greenland, on my first assignment in my new job as geophysicist at the Danish Meteorological Institute, setting up a chain of magnetometer stations on the West Coast."[7]

Dr. Friis-Christensen remembers lying in his tent and "watching the ink pens of my recorder going so wild that they nearly tore the paper chart apart—we had no digital recording at that time—and I wondered whether such big events could also have an influence in the lower atmosphere, on weather and climate. . . . That storm cut off my contact to the outside world for nine days—all radio communication was blacked out—so I had lots of time to reflect on the enormity of the forces at play."[8]

Dr. Friis-Christensen would soon discover he had a soul mate in his reflections, his mentor and a division head at the institute, Knud Lassen, a pioneer in research into the aurora borealis. They followed developments in the field, even gave lectures on the subject, which was then topical, although not for the reasons we're familiar with today: in the mid-1970s, climate scientists feared global cooling. Yet for both scientists, the interest was more a hobby than a formal area of study—until 1989, when Dr. Lassen, sixty-eight years old and nearing retirement, decided to cap his career by pursuing a hunch they had long held. Dr. Friis-Christensen needed no persuading to join him on his quest.

Two years later, they published a path-breaking study in *Science* showing a startling correlation between global temperature and the sunspot cycle.[9] That was not the surprise however. "In 1991, we found that it was not, what everybody believed at that time, the sunspot number itself that was the best proxy for the responsible agent on the Sun. The sunspot cycle length was probably a better descriptor, at least during the last 100 years."[10] Did this mean that sunspot cycle length was "causing" global temperature change? Not at all. The significance was that by finding that cycle length rather than sunspot quantity correlated better to temperature change, they might develop some insight into the physical mechanism relating the two phenomena. "For a physicist a pure correlation is not in itself interesting. The real value of a correlation is if it can indicate where to search for the physical mechanism."[11]

The paper was not immediately controversial in the way Friis-Christensen's later work would be. Global cooling had receded from public memory and global warming was not yet a hot political topic. But it marked the beginning of a wave of related research by solar scientists seeking to learn the mechanisms through which solar activity may influence climate.

When the IPCC was created, Dr. Friis-Christensen hoped its work would spur interest in the Sun's influence on climate change. To participate in the IPCC's quest for answers, he traveled to its January 1992 meeting in Guangzhou, China, as part of the Danish delegation. By then, he had succeeded Dr. Lassen to become head of the institute's geophysics division. As he explained to me, "It is important to understand that my own motivation for doing this research was not to contribute to the global warming discussion. In 1991, this was only in the beginning phase. I wanted to understand what it was in the Sun that could change climate in the Earth, since so many observations indicated a connection."[12]

To his surprise, the IPCC refused to consider the Sun's influence on Earth's climate as a topic worthy of investigation. The IPCC conceived its task as investigating manmade causes of climate change.

Far from being invited to play a role, he found himself rather quickly designated as an enemy, with advocates of the CO2 theory ferociously attacking his credibility. His 1991 study had errors, his detractors stated. A follow-up study in 1995 only made it worse, others chimed in. He fabricated data, people whispered. He made mistakes in his arithmetic. This sort of bickering is not unheard of in scientific circles. But the attacks on Friis-Christensen and colleagues such as Henrik Svensmark—another Dane playing a leading role in solar-climate research—were especially venomous. Quite unintentionally, the solar scientists found themselves in the middle of a political battle royal, their work being used by one side and abused by the other. And the use and abuse mostly focused on a quite predictable issue: the correlations.

Correlations. Laymen love them. Politicians love them. Al Gore and scaremongers of all varieties love them. And so do his critics. Put a big chart up on the wall that shows temperature and CO_2 moving beautifully together, and most people would leave convinced that the oceans are about to boil. Put up another chart cutting the time frames a little differently, zooming in or zooming out, and the correlation disappears. Most might leave *that* meeting convinced that global warming is hyped up, or even a "green" plot.

But for Friis-Christensen and Svensmark, indeed for any honest scientist, the correlation is "interesting, but only as a hint, because climate is not determined by just one thing, and you will never expect a perfect correlation in nature."[13] No matter how powerful a given climate factor is in itself, at any given time, it can be overwhelmed by other factors or inspire negative feedbacks.

Correlations alone cannot sort out the complex of causality. The CO_2 theory is taken seriously only because its (quite rough) correlations to temperature change are backed up by an undeniable physical mechanism. This does not mean that at any given moment CO_2 is necessarily the most powerful influence on temperature. One reason the correlations between CO_2 and temperature don't always look so strong is that other factors may be overwhelming the effect of CO_2 at that moment. The same applies to the Sun.

In Friis-Christensen and Lassen's 1991 paper, they used data up through 1985, the latest then available, which showed a strong correlation between temperature and solar activity. Starting almost immediately thereafter, this correlation broke down: temperatures continued to rise while solar activity fell. Friis-Christensen and essentially the whole solar-climate community acknowledged this quite clearly in subsequent papers. For instance, in a 2000 paper Lassen and Friis-Christensen specifically noted that it was possible that CO_2 effects were for the moment crowding out solar effects.[14]

But these imperfect correlations did not for a moment suggest to the solar scientists that they should discontinue their work—because the point of their work had never been to establish a correlation. The point of their work was to search out the physical processes various correlations might be suggesting. By this time Friis-Christensen and colleagues such as Svensmark and Nigel Marsh were already working on a candidate physical mechanism: the influence of cosmic radiation on cloud cover, perhaps through ionization effects. The magnetic field of the Sun, which varies in strength over time, deflects cosmic rays away from our solar system. The stronger the magnetic field, the fewer the cosmic rays reaching Earth. If cloud cover varies with cosmic rays, then climate could vary with the solar magnetic field.

Clouds, however, are not well understood. Clouds come in all shapes and sizes. Their lifetime varies from minutes to months. They can be less than a square mile in area, or they can ring the Earth. There are dark clouds and bright clouds. There are clouds that on balance capture more of the Sun's radiation than they reflect and clouds that on balance reflect back out to space more than they capture. Clouds can enhance the greenhouse effect and clouds can reduce it. And we know precious little about any of them. In fact, one of the most persistent complaints one hears about climate models is that they can't handle clouds. And historical cloud data is even scantier than CO_2 or temperature data, or even solar magnetic field or irradiance data. In modern times, we have an "instrumental" temperature record starting from 1860. But cloud cover cannot be tracked effectively from the ground. Practically speaking, the "instrumental record" on cloud coverage begins with the satellite era. As for "pre-instrumental" data of the sort we have for CO_2 and temperature, there is essentially none. Clouds don't leave a geological footprint that we can read.

At a 1996 conference in Birmingham, England, Friis-Christensen in an invited paper included some results from a recently completed paper of Svensmark and Friis-Christensen tentatively suggesting that cosmic ray flux, regular variations in the Sun's magnetic field, might be influencing cloud formation and thus temperature.[15] If the IPCC and the CO_2 theorists had been hostile before, their reaction to the 1996 paper was explosive. As Svensmark told me, "To my surprise there was a very

THE CV OF A DENIER
Dr. Henrik Svensmark

Henrik Svensmark is director of the Center for Sun-Climate Research at the Danish National Space Center. Previously, Dr. Svensmark was head of its Sun-Climate group. He has held post-doctoral positions in physics at University California Berkeley, Nordic Institute of Theoretical Physics, and the Niels Bohr Institute. In 1997, Dr. Svensmark received the Knud Hojgaard Anniversary Research Prize, and in 2001, he was the recipient of the Energy-E2 Research Prize.

strong reaction to the idea that cosmic rays and solar activity could influence the Earth's climate at that time. It was announced at a scientific meeting in Birmingham in 1996. The reaction was very strong from the media and it made headlines in Denmark. It was unexpected for me that the reaction was so strong, and I remember coming back to Copenhagen seeing a headline saying, 'Strong UN critique of Danish researchers.'"[16] Bert Bolin, then chairman of the United Nations IPCC, castigated them in the press, saying, "I find the move from this pair scientifically extremely naïve and irresponsible."[17] They were denounced for casting doubt on the greenhouse theory, "though we did not say at that time that there was no CO_2 effect, just that the Sun is also important."[18]

Eventually the Danes saw that the correlation between the Sun's magnetic field and total cloud cover broke down also. Far from a failure, this turned out to be a breakthrough, because it led them to a more powerful (though never perfect!) correlation between cosmic radiation and low cloud cover. Ultimately, the focus on low clouds would provide a strong hint about a physical mechanism.

Low altitude clouds are significant because they especially shield the Earth from the Sun to keep us cool. Low cloud cover can vary by 2% in five years, affecting the Earth's surface by as much as 1.2 watts per square meter during that same period. "That figure can be compared with about 1.4 watts per square meter estimated by the Intergovernmental Panel on Climate Change for the greenhouse effect of all the increase in carbon dioxide in the air since the Industrial Revolution,"[19] Dr. Svensmark explains.

In the tale as told by their more vitriolic critics the back and forth in the Danes's research—the correlations that hinted at the Sun's role but never perfectly, the papers that were dutifully updated when new data seemed to undermine the old—was portrayed not as the ordinary progress of science but as the obfuscations and double-talk of charlatans. But as Popper taught us, science proceeds by falsification. As Friis-Christensen told me, "True, the correlation with total cloud cover stopped. What does a real physicist do? Examines the options: either the hypothesis is wrong, or it has to be refined. Again, newer and better data allowed to go to more detailed altitude resolution, which showed that the correlation was with the low clouds. . . . [F]or the first period examined it just happened that the total cloud cover variation did show the same variations because the high and middle clouds together showed no variation during that period. This is precisely how science works. Our objective never was just to fight the point of view of manmade global warming."[20]

Meanwhile, the preliminary research of the Sun scientists began to be cited by politicians, lobbyists, and journalists who opposed the doomsayer camp's calls for immediate drastic cutbacks in hydrocarbon emissions. On the other side, the same thing was happening. The hockey stick and other headline-making exhibits were being used to "prove" the CO2 hypothesis and to support drastic cutbacks in emissions. Al Gore made his documentary, and a British producer, Martin Durkin, responded with *The Great Global Warming Swindle*, his no-holds-barred refutation.

Perhaps unwisely, Friis-Christensen agreed to be interviewed for the Durkin documentary. To his later dismay, Durkin highlighted the early cloud-cover correlation, without telling his audience that the correlations had ended. "From the movie makers' side it is probably felt that the correlation is perhaps the easiest way to explain to the general audience how the mechanism works,"[21] Friis-Christensen muses. And for the same reason, it is the confounding evidence in the correlations that critics harp on. "And from the other side, the main target for attack is precisely the suggestive illustrations and not the physical mechanism behind them."[22]

The physical mechanism. That really was the problem. Almost 15 years after Lassen and Friis-Christensen had published their first paper suggesting a link between the Sun and recent global warming, the Danes still had no reproducible physical demonstration of how it might happen. That was about to change.

CHAPTER TEN

SKY and CLOUD

Sami Solanki, Jasper Kirkby,
Henrik Svensmark, Nir Shaviv

The Danes were never alone in their suspicions that changes in the Sun could affect the Earth's climate. In Germany, for instance, Dr. Sami Solanki of the Max Planck Institute for Solar System Research in Germany had also observed powerful correlations between various measures of solar activity and climate. In 2004, he led a team of scientists that quantitatively reconstructed the Sun's activity since the last Ice Age, some 11,400 years ago, mapping this activity against temperature records. They concluded that the Earth hasn't been as hot as it is now in 8,000 years and that much of the increase correlates with increased brightness of the Sun.

The 19th and 20th centuries he believes especially noteworthy. "The Sun is in a changed state. It is brighter than it was a few hundred years ago and this brightening started relatively recently—in the last 100 to 150 years,"[1] he says. "The Sun has been at its strongest over the past 60 years and may now be affecting global temperatures."[2]

Solanki's 2004 paper marked a pivotal point in the climate-change debate, invigorating research into solar factors involved in climate change. The furious denunciations against Friis-Christensen began to ring rather hollow when Max Planck Institute (among others) joined the fray.

Like the Danes, however, Solanki is far from a partisan denier. His research also showed that the various correlations between solar behavior and temperature abruptly broke down in the 1980s. (See figure 1.) He thinks it is quite likely that CO_2 explains a portion of warming in recent decades. He insists, however, that much of the warming of the past 150 years cannot be explained by CO_2 emissions and must have another source, with the Sun as an excellent candidate. And while he does not think that cosmic ray/low cloud cover hypothesis will turn out to be the only relevant solar factor, he does say that "the mechanism is just too beautiful to ignore."[3]

THE CV OF A DENIER
Sami Solanki

Sami Solanki is director and scientific member at the Max Planck Institute for Solar System Research in Germany. Previously, he was appointed professor of astronomy at the University of Oulu in Finland in 1998 and Minnaert professor at the University of Utrecht in the Netherlands in 1999. Among his research interests are solar physics, the physics of cool stars, radiative transfer, and astronomical tests of theories of gravity. Dr. Solanki obtained his doctorate from the ETH (Eidgenössische Technische Hochschule) in Zurich in 1987. His Web site is www.mps.mpg.de/homes/solanki.

Correlations between cosmic ray flux and global temperature are suggestive, but inconclusive

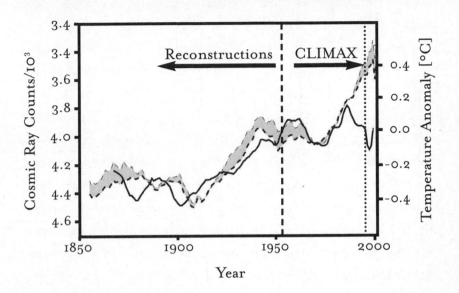

Figure 1. Cosmic ray flux (solid line) compared to two different temperature reconstructions (from Krivova & Solanki, 2003). In the mid-1980s cosmic ray flux and temperature diverge after correlating very strongly for more than 100 years. *Adapted from* N. A. Krivova & S. K. Solanki, "Solar Total and Spectral Irradiance: Modeling and A Possible Impact on Climate," abstract, *ESA* SP 535 (2003), p. 281. See: http://www.mps.mpg.de/projects/sun-climate/papers/iscs2003.pdf.

Surprisingly, some of the strongest evidence of the existence of this "beautiful" mechanism—evidence that cosmic rays could affect climate—did not come from the Sun or the sky, but from stone. From meteorites.

★   ★   ★

Nir Shaviv is a young man. But the subject of his original research is very old. Hundreds of millions of years old. An astrophysicist by training, he plunged into the study of iron meteorites that had spent the past several hundred million years roaming interplanetary space before striking Earth. He noticed a curious thing. There was a very large variation in the amount of cosmic ray damage sustained by meteorites. For tens of millions of years, the cosmic ray damage would be quite large. Then for similar periods of time the cosmic ray damage would be relatively slight. Then it would start up again and heavy damage would again persist for tens of millions of years.

Did the time periods mean anything? Was there an interesting correlation? Yes, as it turned out. The solar system is not stationary relative to our galaxy, the Milky Way. We revolve around the center of the galaxy at a speed of about half a million miles per hour. During this journey, we alternate between passing through one of the galaxy's spiral arms, densely packed with massive stars and dust, then across vast areas devoid of massive stars, and then on into the next spiral arm. What Shaviv found, as published in the prestigious journal *Physical Review Letters*, was that meteorites sustained at least 2.5 times as much cosmic ray damage while traveling through the arms of the galaxy as they did in the relatively empty space between arms. This makes sense, since the supernovae that are the source of cosmic rays tend to be found in the spiral arms.[4]

Shaviv was not doing climate research at the time. But he learned that the Danes and other space scientists in Denmark had raised the possibility of a connection between cosmic rays and cloud formation and thus climate. He quickly saw the possible link to his own work.

"If this [connection exists], then one should expect climatic variations while we roam the galaxy. . . . [E]ach time we cross a galactic arm [once every 135 ± 25 million years], we should expect a colder climate."[5]

After reconstructing the temperature of Earth over the past 550 million years and matching that history against a history of cosmic ray flux, his hunch was borne out. The cosmic ray flux was "in sync with the occurrence of ice-age epochs on Earth."[6] How could this work? If the cosmic ray/cloud hypothesis were correct, the increase in cosmic ray density when the Earth is within a spiral arm "could increase Earth's cloud cover by as much as 15%, altering global temperatures by five to 10 degrees. . . ."[7] This is "more than enough to change the state of Earth from a hothouse, with temperate climates extending to the polar regions, to an icehouse, with ice caps on its poles, as Earth is today."[8] Shaviv argues that these cosmic ray fluctuations explain more than two-thirds of Earth's temperature variance. "[C]osmic rays undoubtedly affect climate, and on geological time scales are the most dominant climate driver."[9]

Though Earth isn't passing into or out of a spiral arm now, Shaviv's research provides "evidence of the extent to which cosmic forces influence Earth's climate."[10] And there are other factors that affect cosmic ray intensity on smaller scales, like changes in the intensity of the Sun's magnetic field. The more powerful the field, the more it acts as a shield steering cosmic rays away from the Earth.

The Danes and others quickly saw the relevance of Shaviv's research and began citing it in their own papers. Svensmark plunged into Shaviv's arena and and began his own exploration into a link between "the evolution of the climate and the biosphere on the Earth" and "the evolution of the Milky Way."[11]

Still they did not have a reproducible physical mechanism. But in 2000, in a paper published in *Space Science Review*, Svensmark and colleague Nigel Marsh had cautiously proposed a hypothesis based on the ionizaton of atmospheric particles by cosmic rays. "The ionizing potential of Earth-bound cosmic rays are modulated by the state of the heliosphere, while clouds play an important role in the Earth's radiation budget through trapping outgoing radiation and reflecting incoming radiation."[12] (The heliosphere is a sort of cosmic umbrella cast over the solar system by the Sun's magnetic field.) "If a physical link between these two features can be established, it would provide a mechanism linking solar activity and Earth's climate."[13]

The link would most likely be focused on low clouds because of their temperature and composition. "Low clouds are warm (>273K) and therefore consist of liquid water droplets. At typical atmospheric supersaturations . . . a liquid cloud drop will only form in the presence of an aerosol, which acts as a condensation site. . . ."[14] Moreover, a rather small change in aerosol production could be enough to create a big change in low cloud cover. "Observations of local aerosol increases in low cloud due to ship exhaust indicate that a small perturbation in atmospheric aerosol can

The Danish National Space Center on the Sources of Climate Change

'Climate is subject to influences by both natural and human forces, including greenhouse gases, aerosols, solar activity, and land use change. . . .

"[T]he varying activity of the Sun is indeed the largest and most systematic contributor to natural climate variations. The effect goes through solar modulation of the cosmic radiation, which affects the formation of aerosols and thereby also the formation of clouds. . . .

"Solar activity has been exceptionally high in the 20th century compared to the last 400 years and possibly compared to the past 8,000 years. When solar activity is high, the flux of galactic cosmic rays is reduced due to increased magnetic shielding by the have a major impact on low cloud radiative properties. Thus, a moderate influence on atmospheric aerosol distributions from cosmic ray ionization would have a strong influence on the Earth's radiation budget."[15]

Not until six years later would they be able to test the proposed mechanism in a lab. In 2006, Svensmark assembled a team at the Danish National Space Center to undertake an elaborate experiment in a reaction chamber the size of a small room. Dubbed SKY (Danish for "cloud"), the experiment mimicked salient features of the chemistry of the lower atmosphere, adding ultraviolet rays to mimic the actions of the Sun. Naturally occurring cosmic rays were filtered in through the ceiling.

What they found left them agape: a vast number of floating microscopic droplets soon filled the reaction chamber. These were

Sun. The cosmic rays may influence Earth's climate through formation of low-lying clouds.

"Cosmic rays ionize the atmosphere and an experiment performed at the Danish National Space Center has found that the production of aerosols in a sample atmosphere with condensable gases (such as sulfuric acid and water vapor) depends on the amount of ionization. Since aerosols work as precursors for formation of cloud droplets, this is an indication that cosmic rays affect climate.

"Climate models only include the effects of the small variations in the direct solar radiation (infrared, visible, and UV). The effects of cosmic rays on clouds are not included in models, and the models do a rather poor job of simulating clouds in the present climate. Since cloud feedbacks are a large source of uncertainty, this is a reason for concern when viewing climate model predictions."

From the Center for Sun-Climate Research at the Danish National Space Center, Web site, http://www.spacecenter.dk/research/sun-climate

ultra-small clusters of sulfuric acid and water molecules—the building blocks for cloud condensation nuclei—that had been catalyzed by the electrons released by cosmic rays. They had expected some effect. The surprise was that the electrons acted as catalysts—each causing not one but several reactions before being lost to the environment. This strengthened their notion that a relatively small change in cosmic radiation could have a significant effect on climate.

★　★　★

"We were amazed by the speed and efficiency with which the electrons do their work," Dr. Svensmark remarked. For the first time ever, researchers had experimentally identified a causal mechanism by which cosmic rays can facilitate the production of clouds in Earth's atmosphere. "This is a completely new result within climate science."[16]

Whenever I read a comment from some journalist or politician about how the solar theorists are cranks pursuing crackpot theories in backwater institutions, I smile and think about CERN. Based in Geneva, CERN is the European Organization for Nuclear Research, a 50-year-old institution, originally founded by 12 countries and now counting 20 country-members. It services 6,500 particle physicists—half of the world's total—in 500 institutes and universities around the world. It is building the $2.4-billion Large Hadron Collider (LHC), the world's most powerful particle accelerator. And as the home to Jasper Kirkby's CLOUD project, it is now the center stage for research into solar theories of climate change and host to one of the most significant climate-change experiments to be proposed in our time.

CLOUD almost didn't happen. Dr. Kirkby, not unlike Friis-Christensen and Svensmark, is a superb scientist but clueless as a politician. In 1998, anticipating he'd be leading a path-breaking

THE CV OF A DENIER
Jasper Kirkby

Jasper Kirkby is author or coauthor of some 250 scientific publications. The Royal Swedish Academy of Sciences describes Kirkby as "an experimental particle physicist at CERN, Switzerland. After completing his degrees at Oxford and London, he spent 12 years at Stanford before joining CERN in 1984. While at CERN he originated the idea for a new accelerator known as the Tau-Charm Factory, which is now under construction as BEPC II in Beijing. He has originated and led several large experiments at accelerators, including DELCO at SPEAR and PEP; FAST at PSI; and the CLOUD experiment at CERN."[17]

experiment into the role of cosmic rays on cloud formation, he let his enthusiasm get the best of him. He actually told some journalists he thought that cosmic ray flux might "account for somewhere between a half and the whole of the increase in the Earth's temperature that we have seen in the last century."[18] Knowing how savagely Svensmark and Friis-Christensen had been attacked for even raising the possibility of a cosmic ray connection, he really should have known better.

Kirkby's proposed experiment, which had heretofore flown safely beneath the global warming radar, suddenly became infamous. He was condemned for minimizing the role of human beings in global warming. He was attacked for supposedly playing into the hands of oil-industry lobbyists. And the funding he had been assured of was suddenly put on ice. Kirkby was stunned. CERN had never before seemed so vulnerable to the whims of government funders.

It took ten years for Kirkby's experiment to recover from the blow. But now CLOUD is actually underway. The purpose of CLOUD (Cosmics Leaving Outdoor Droplets) is to establish the mechanisms through which cosmic rays can influence the formation of clouds and thus the climate. The experiment will use a high-energy particle beam from a CERN accelerator to closely duplicate cosmic rays found in the atmosphere. (The SKY experiment used naturally occurring cosmic rays.)

As a CERN statement on the experiment puts it, "By studying the micro-physical processes at work when cosmic rays hit the atmosphere, we can begin to understand more fully the connection between cosmic rays and cloud cover."[19] This is crucial because "[c]louds exert a strong influence on the Earth's energy

Comparing CLOUD and SKY

The European Organization for Nuclear Research, CERN, is creating an atmospheric research facility at its particle physics laboratory in Geneva. Called CLOUD, it will consist of a special cloud chamber exposed to pulses of high-energy particles from one of CERN's particle accelerators, the Proton Synchrotron. Conditions prevailing in the Earth's atmosphere will be re-created in CLOUD, and the incoming particles will simulate the action of cosmic rays. An elaborate set of instruments will trace the physical and chemical effects of the particle pulses.

The project was first proposed by Jasper Kirkby of CERN in 1998, in response to the discovery in Copenhagen of an apparent link between cosmic rays and clouds. The name CLOUD is an acronym for "cosmics leaving outdoor droplets." More than fifty atmospheric scientists, solar-terrestrial physicists, and particle physicists from 17 institutes in Europe and the United States joined Kirkby's team, including Henrik Svensmark and colleagues balance, and changes of only a few per cent have an important effect on the climate."[20]

To accomplish all this, Dr. Kirkby has assembled a dream team of atmospheric physicists, solar physicists, and cosmic ray and particle physicists from 18 institutes around the world including the California Institute of Technology; the Danish National Space Center; the Lebedev Physical Institute, Russia; the Leibniz Institute for Tropospheric Research, Leipzig, Germany; the Max-Planck Institute for Nuclear Physics, Germany; the Rutherford Appleton Laboratory, United Kingdom; and more.[21]

While the world of particle physics is eagerly awaiting the CERN/CLOUD results, the world of climate science, enthralled by the money and political muscle behind the doom-

at the Danish Space Research Institute (now called the Danish National Space Center).

Financial problems at CERN left the project "on ice" until its eventual approval in March 2006. By then, the simpler SKY experiment in Copenhagen was beginning to give results. The multinational team decided that the first use of the experimental site should be a rerun of SKY, in the autumn of 2006. The Danish team has built a duplicate of the apparatus, SKY-2, which was tested successfully at CERN and provided important inputs for the design of the CLOUD facility. Other scientists from the CLOUD team contributed additional instruments.

The main cloud chamber for the CLOUD facility is expected to begin operating in 2010. Apart from the use of controllable accelerated particles rather than natural cosmic rays, and more extensive instrumentation, the main difference from the SKY experiment is that CLOUD will be able to use air at low pressures and low temperatures. It will thus reproduce conditions high in the atmosphere while SKY explores cosmic ray action only in warm, dense air close to the ground.[23]

sayer account of global warming, is all but ignoring the potential of CLOUD to advance our understanding of climate. The 2007 report of the IPCC ranked the Sun as an all-but-irrelevant factor in climate change and does not take into account the cosmic ray theory. As for Dr. Kirkby, he, perhaps wisely, refused to be interviewed for either my column or this book. From the little he does say, he could almost be part of the "consensus." He knows what happens to those who speak out against the party line and downplays the significance of what CLOUD may find: "If there really is an effect, then it would simply be part of the climate-change cocktail,"[22] a perhaps less naïve, more politic Dr. Kirkby now states.

CHAPTER ELEVEN

Cycles Within Cycles

Habibullo Abdussamatov, George Kukla, Rhodes Fairbridge

The cosmic ray/cloud connection is complex and intriguing, but there may be a less complex and more intuitive way to tell that the Sun is responsible for global warming. Consider this: Mars has been warming too. Its polar ice cap is shrinking, deep gullies in its landscape are now laid bare, and the Martian climate is the warmest it has been in decades or centuries. Mars may be "just coming out of an ice age,"[1] NASA scientist William Feldman speculated after the agency's Mars Odyssey completed its first Martian year of data collection. "In some low-latitude areas, the ice has already dissipated."[2]

To Dr. Habibullo Abdussamatov, the head of the Space Research Laboratory at St. Petersburg's Pulkovo Astronomical Observatory, the evidence from Mars destroys in terms the layman can understand the notion that humans are responsible for warming Earth. "Mars has global warming, but without a greenhouse and without the participation of Martians,"[3] he states matter-of-factly.

Dr. Abdussamatov, at the pinnacle of Russia's space-oriented scientific establishment, is also one of the world's most eminent critics of the notion that manmade CO_2 is driving global warming. He argues that these "parallel global warmings—observed simultaneously on Mars and on Earth—can only be a straight-line consequence of the effect of the one same factor: a long-time change in solar irradiance."[4]

Abdussamatov's hypotheses are taken so seriously in Russia that they have been made a focus of Russian experiments on the International Space Station. His work revolves around a complex of cyclical variations in total solar irradiance (TSI) that, he believes, relates solar activity to climate change.

The Sun's total irradiance varies according to larger and smaller cycles. There is the "solar cycle" of sunspot and solar irradiance activity, typically running about 11 years, which characteristically comprises a small variance in total solar irradiance, usually too small to be strongly correlated to climate change. But there's also a larger cycle, running about 200 years. Abdussamatov argues that "the predominating factor influencing climate changes (geophysical effects lasting for decades) is alterations in a general two-century-long cycle of solar activity defined primarily by corresponding changes in TSI."[5]

However, there is an additional complexity: the change in Earth's climate lags the change in the Sun, largely because of the effects of the oceans, which have an enormous capacity to absorb heat. In the same way that a cast-iron frying pan heats up when a stove is turned on and then continues to give off heat well after the stove has been turned off, the oceans absorb heat during times of increasing irradiance and release it when irradiance subsides. The release of heat from the oceans continues for 15 to 20 years after the irradiance has subsided. In this way, the oceans moderate short-term temperature variations within an 11-year solar cycle, diminishing the Sun's apparent influence on climate.

The oceans—whose upper regions contain many times the CO_2 in Earth's atmosphere—are also responsible for confusing climate scientists about the role that CO_2 plays in the warming of the planet. "If the temperature of the ocean rises even a little, gigantic amounts of CO_2 are released into the atmosphere through the evaporation of water,"[6] explains Dr. Abdussamatov. "It is no secret that increased solar irradiance warms Earth's oceans, which then triggers the emission of large amounts of carbon dioxide into the atmosphere. So the common view that man's industrial activity is a deciding factor in global warming has emerged from a misinterpretation of cause-and-effect relations."[7]

The Sun directly accounts for about half the warming that we have seen on Earth in the 20th century, Dr. Abdussamatov believes. "The other half is caused by the natural greenhouse effect and by the natural variations in the albedo of the Earth's surface," he states, "but almost none of it stems from a manmade greenhouse effect. The natural increase of water vapor, the natural increase of CO_2, and decrease in the albedo of the Earth's surface from the warming in the 20th-century dominated global warming."[8]

Furthermore, Abdussamatov believes that this recent global warming will be short-lived and that we are actually on the brink of a global cooling, likely a severe one. He argues that Earth has hit its temperature ceiling, demonstrated by cooling that is occurring on the upper layers of the world's oceans. Solar irradiance has begun to fall, ushering in a protracted cooling period beginning in 2012–2015. The depth of the decline in solar irradiance reaching Earth will occur around 2041 (plus or minus 11 years, he estimates) and "will inevitably lead to a deep freeze around 2055–60"[9] lasting some 50 years, after which temperatures will go up again. "We continue to bask in the remains of heat that the planet accumulated over the 20th century."[10]

Figures 1 and 2 help illustrate his argument. Figure 1 shows the last few 11-year cycles. There is marked variation within the cycles. The crucial point, however, says Abdussamatov, is that the long-term trend has already begun to head down—even as we experienced continued warming.

It is the thermal inertia of the Earth that has prolonged the most recent global warming, while the impact of recent declines in

A cooling in our future?
Solar activity, though volatile, is trending down

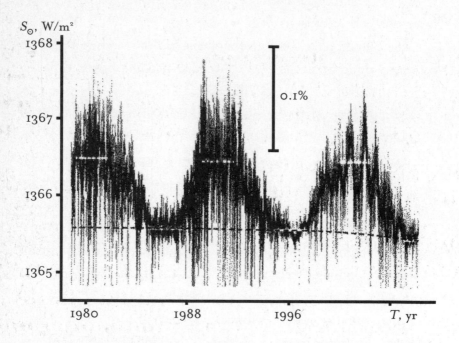

Figure 1. Recent cycles of solar activity with their long-term trend line. (Dashed line.) Volatility within the cycles can be confusing. But, says, Abdussamatov, the long-term trend has already begun to head down, even as we experience continued warming due to climatic inertia, especially from the oceans. *Adapted from* Kh. I. Abdusamatov, "Optimal Prediction of the Peak of the Next 11-Year Activity Cycle and of the Peaks of Several Succeeding Cycles on the Basis of Long-Term Variations in the Solar Radius or Solar Constant," abstract, *Kinematics and Physics of Celestial Bodies* 23, no. 3 (2007), p. 98. [© Allerton Press, Inc.]; [Original Russian Text © Kh.I. Abdusamatov, published in *Kinematika i Fizika Nebesnykh* Tel 23, no. 3 (2007), pp. 141–147.]

Solar activity is consistent with global temperature change, assuming climatic "inertia," driven by oceans

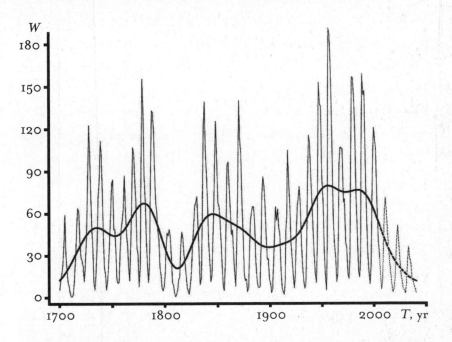

Figure 2. Solar activity from 1700 with Abdussamatov's projections out through 2040, including trend line. The peak of activity roughly coincides with 20th-century warming after accounting for the lag resulting from the "inertia" of Earth's climate, largely driven by the oceans. *Adapted from* Kh. I. Abdusamatov, "Optimal Prediction of the Peak of the Next 11-Year Activity Cycle and of the Peaks of Several Succeeding Cycles on the Basis of Long-Term Variations in the Solar Radius or Solar Constant," abstract, *Kinematics and Physics of Celestial Bodies* 23, no. 3 (2007) p. 99. [© Allerton Press, Inc.]; [Original Russian Text © Kh.I. Abdusamatov, published in *Kinematika i Fizika Nebesnykh* Tel 23, no. 3 (2007), pp. 141–147.]

total solar irradiance cannot yet be felt. "Global cooling will come relatively soon as the planet having received increased solar energy over almost all of the 20th century now gradually gives it back."[11] In support of this he cites studies showing "cooling of the upper ocean," which began in 2003–2005 (John M. Lyman, Josh K. Willis, and George C. Johnson, "Recent Cooling of the Upper Ocean,"

2006). Not until the upper ocean has been cooling for six to eight years will we "feel a very slow beginning of global cooling."[12] The Russian and Ukrainian space agencies, under Dr. Abdussamatov's leadership, have launched a joint project to determine the time and extent of the global cooling at mid-century. The project, dubbed Astrometria, has been given priority space-experiment status on the Russian portion of the International Space Station and will marshal the resources of spacecraft manufacturer Energia, several Russian research and production centers, and the main observatory of Ukraine's Academy of Sciences. By late 2010, scientific equipment will have been installed in a space-station module, and by early 2011, Dr. Abdussamatov's space team will be conducting a regular survey of the Sun. With six years of data from the Astrometria project, he hopes to "give a more precise scenario of the upcoming climatic changes on Earth by 2016."[13]

We may not need to wait so long for an update however. Mars should tell us first. "Since there is no ocean on Mars, its thermal inertia is significantly lower. Hence cooling on Mars will start earlier than on the Earth," giving us "confirmation of a future cooling of the Earth."[14]

Abdussamatov's work is not looked upon kindly by the global warming doomsayers, who dismiss him in the same scornful tones they use with the Danes. The most common argument is that Abdussamatov is misinterpreting the undeniable recent warming of Mars. *Nothing to do with the Sun*, say the critics; Mars is warming because of variations in its orbit. Abdussamatov returns that this is impossible because "the form of the orbit and tilt of the axis of both Mars and Earth varies on the time scale of tens thousands of years, and these variations cannot increase the value of solar irradiance of Mars during so short a period of time."[15]

If Abdussamatov and his colleagues are right, the real crisis we

THE CV OF A DENIER
Habibullo Abdussamatov

Habibullo Abdussamatov, born in Samarkand, Uzbekistan, in 1940, graduated from Samarkand University in 1962 as a physicist and a mathematician. He earned his doctorate at Pulkovo Observatory and the University of Leningrad.

He is the head of the space research laboratory of the Russian Academy of Sciences' Pulkovo Observatory and of the International Space Station's Astrometria project, a long-term joint scientific research project of the Russian and Ukranian space agencies.

face is learning how to deal with a century of falling temperatures, during which time we will enter a mini ice age. The glaciers will advance, snow will cover increasing parts of the globe, and greenhouse gases—primarily water vapor but also CO2—will leave the atmosphere to return to the oceans. The activities of humans will be regarded as having no influence on the cooling of the globe, just as their activities will amount to having no influence on its warming. Dr. Abdussamatov concluded, "A global freeze will come about regardless of whether or not industrialized countries put a cap on their greenhouse gas emissions."[16]

Abdussamatov is not alone in his expectation of a coming chill. In the 1970s, a number of respected scientists claimed that the world was threatened by an era of global cooling. Their warnings not only provoked official action from the U.S. government but actually led to the eventual formation of the UN ICCP (International Climate Change Partnership). Prominent among those researchers was George Kukla of Columbia University and the

Czechoslovakian Academy of Sciences. Today, based on new data, Dr. Kukla thinks the warnings he and his colleagues issued 30 years ago were right: the world could soon be entering another ice age. Like Abdussamatov, Kukla stresses that global temperature trends can be misleading, or at least lagging indicators. It is to be expected, he now says, that average global temperature would continue to rise even after the cooling process has begun.

"Only in the last five or six years did I fully realize how tricky is the measure of a global mean temperature,"[17] he explains, referring to the calculation scientists use to determine Earth's surface temperature. In this calculation, the relatively small polar regions have little weight, with the result that "the tropical belt plus the adjacent middle latitudes, which represent almost three-quarters of the globe surface, define the "global warming."[18]

For the last 6,000 years, the tropical and adjacent latitudes have for the most part been slowly warming. But this slow warming, on its own, tells us little. "What seems to be the principal driver of weather patterns and long-term global climates is the temperature difference between the high latitudes, in the polar regions, and the tropical belt,"[19] Dr. Kukla explains. "The higher the temperature gradient, the more water is moving toward the poles by ocean and atmosphere currents and remains there as snow and ice."[20]

"So, funny as it sounds, the glacials, or periods of global cooling, start with a peak high global mean temperature—a crucial element in the transfer of water from warming oceans to the cooling subpolar lands. Only when the resulting ice, which can be understood as a very slow "liquid," flows toward the low latitudes, does the global area-weighted mean surface temperature start to drop. You could say that we enter the global cooling period by a global warming."[21]

Dr. Kukla, in 1972 a member of the Czechoslovakian Academy of Sciences, and a pioneer in the field of astronomical climate forcing, became a central figure in convincing the U.S. government to take the dangers of climate change seriously. In January of that year, he, Robert Matthews, chairman of the Department of Geological Sciences at Brown University, and Murray Mitchell, chief climatologist of the National Oceanic and Atmospheric Administration, convened what would become an historic conference of top European and American investigators at Brown University in Providence, Rhode Island. The working conference's theme: "The Present Interglacial: How and When Will It End?"

Later that year, Drs. Kukla and Matthews highlighted the dangers of the expected global cooling in *Science* magazine and because of the supposed urgency of the matter alerted President Richard Nixon in a joint letter. The conference had reached a consensus, their letter stated, that "a global deterioration of climate, by order of magnitude larger than any hitherto experienced by civilized mankind, is a very real possibility and indeed may be due very soon. The cooling has natural cause and falls within the rank of processes which produced the last ice age."[22]

The White House reacted swiftly to the letter, which predicted "substantially lowered food production"[23] and "extreme weather anomalies,"[24] such as killer frosts and floods, and included a warning that the Soviet Union might already be in the lead in preparing for the climate disturbances to come. By February 1973, the State Department had established a Panel on the Present Interglacial.

Soon, numerous other government agencies were drawn in— the issue was seen to be of paramount importance—and by 1974, a federal government report, "A United States Climate Program," cited evidence of the climate shift, including:

"A killing winter freeze, followed by a severe summer heat wave in the United States."[25]

"Drought in the Soviet Union producing a 12% shortfall in their grain production in 1972, forcing the country to purchase grain abroad, which in turn reduced world grain reserves and helped drive up food prices."[26]

"Collapse of the Peruvian anchovy harvest in late 1972 and early 1973, related to fluctuations in the Pacific Ocean currents and atmospheric circulation, impacted world supplies of fertilizer, the soybean market, and prices of other protein feed stocks."[27]

"The anomalously low precipitation in the U.S. Pacific Northwest during the winter of 1972–73 depleted water-reservoir storage by an amount equivalent to an amount of water required to generate more than 7% of the electric energy for the region."[28]

By 1975, the *National Climate Program Act of 1975*, the first of numerous such bills, was introduced to establish a coordinated national program of climate research, monitoring, prediction, and contingency-planning analysis. Much congressional testimony spoke of the inadequacy of climate research and the need for preparedness. Meanwhile, the failure of the Soviet Union's wheat crop (and a subsequent high-profile U.S. wheat deal), the severe winter of 1976 to 1977, and El Niño's influence on climate became dinner-table talk. In September 1979, President Jimmy Carter signed the *National Climate Program Act* into law, in aid of predicting future climate and combating global cooling. That act has now been enlisted in the effort to counter global warming.

Many today speak with derision of the 1970s global cooling scare, seeing it as a cautionary false alarm. Others see it as an embarrassment—*Newsweek* magazine, which published a 1975 article entitled "The Cooling World," even corrected the record with a 2006 follow-up to its 1975 article, arguing that scientists who warn about global warming now have it right.

Dr. Kukla sees things differently. Although the 1975 article

THE CV OF A DENIER
Dr. George Kukla

George Kukla, special research scientist at the Lamont-Doherty Earth Observatory of Columbia University, New York, is an expert in the study of solar forcing of climate changes. He was the lead author of the scientific papers that supported the Milutin Milankovic's theoretical explanation of glacial cycles by investigating the radiometrically dated stratigraphy in the deep-sea sediments around the world. In the cores are the clear imprints of Milankovic's proposed cycles. In his papers he wrote, "We are certain now that the changes in the Earth's orbital geometry caused the ice ages. The evidence is so strong that other explanations must now be discarded or modified."[29] Prior to joining Columbia in 1971, Dr. Kukla had published landmark studies in Czechoslovakia, where he was a member of the Czechoslovakian Academy of Sciences. He is the 2003 recipient of the European Geosciences Union's Milutin Milankovic Medal.

indicated that the cooling trend would be continuous, scientists knew otherwise. "None of us expected uninterrupted continuation of the trend,"[30] he states. Moreover, according to Dr. Kukla's recently published evidence, a period of "global warming" always precedes an ice age.

For millions of years, the geologic record shows, Earth has experienced an ongoing cycle of ice ages, each typically lasting about 100,000 years, and each punctuated by briefer, warmer periods called interglacials, such as the one we are now in. The current period of global warming actually constitutes additional indication of the ice age to come.

Furthermore, this ongoing cycle, he says, closely matches cyclic variations in Earth's orbit around the Sun. "I feel we're on pretty solid ground in interpreting the orbit around the Sun as the primary driving force behind the ice-age glaciation. The relationship is just too clear and consistent to allow reasonable doubt,"[31] Dr. Kukla said. "It's either that, or climate drives orbit, and that just doesn't make sense."[32]

The idea that orbital cycles could affect the Sun is not new. Rhodes Fairbridge of New York's Columbia University, a giant in science over much of the last century, has long offered a theory that links solar behavior to planetary orbits.

Fairbridge is a legend in the field whose accomplishments are perhaps unsurpassed for their breadth, depth, and volume. This one man authored or coauthored 100 scientific books and more than a thousand scientific papers, edited the *Benchmarks in Geology* series (more than 90 volumes in print), and was general editor of the *Fairbridge Encyclopedias of Earth Sciences*. He edited eight major encyclopedias of specialized scientific papers in the atmospheric sciences and astrogeology; geomorphology; geochemistry and the earth sciences; geology; sedimentology; paleontology; oceanography; and, not least, climatology.

Fairbridge started with sunspots. Scientists have realized for more than two centuries that changes in sunspots can be correlated with the climate of Earth, the ice ages, and periods of great warming. But what, Dr. Fairbridge wondered, causes these changes in our Sun? His theory, uncovered with the help of NASA and the Jet Propulsion Laboratories, is that sunspot activity is affected by the solar system's center of gravity.

At times, the Sun is at the solar system's center of gravity. But most often this is not the case—planets align to one side or another of the Sun, shifting the system's center of gravity in that direction. Jupiter, the planet with by far the largest mass, has the most influence. When Uranus, Neptune, and especially Saturn—the next largest planet—join Jupiter on the same side of the solar system, the solar system's center of gravity shifts well beyond the Sun. As a result, the center of the mass of the Sun—the barycenter—does not correspond to the barycenter of the system as a whole. (See figure 3.)

Depending on the position of the planets, the barycenter for the system as a whole can be as much as one solar diameter outside the Sun.[33]

Does the solar system's shifting center of gravity affect solar output?

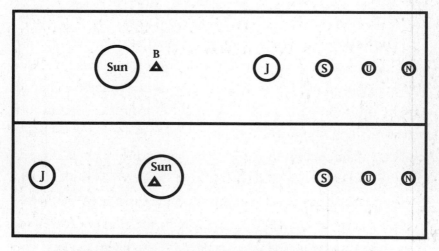

RELATIONSHIP OF THE SUN AND BARYCENTER OF SOLAR SYSTEM

Figure 3. The revolution of the planets about the Sun, especially Jupiter, can cause the center of mass (barycenter) **B** of the Solar System to move from a position within the body of the Sun to a point outside it. Fairbridge hypothesized that the resulting changes in orbital angular velocity of the Sun will cause variations in solar output, affecting climate on the Earth. *Adapted from* Rhodes W. Fairbridge, "The 'Solar Jerk,' The King-Hele Cycle, and the Challenge to Climate Science," Climate and Keplerian Planetary Dynamics Web site (first published in *21st Century Science and Technology*), http://www.crawfordperspectives.com/Fairbridge-ClimateandKeplerianPlanetaryDynamics.htm.

THE CV OF A DENIER
Rhodes Fairbridge

R hodes Fairbridge, an early expert on climate change, was a professor of geology at Columbia University. He received an undergraduate degree from Queen's University in Ontario and a master's degree from Oxford. He was awarded a doctorate of science from the University of Western Australia in 1944, at the age of 30, bypassing the usual Ph.D. prerequisite. During World War II, Dr. Fairbridge also served with the RAAF in General MacArthur's headquarters as deputy director of intelligence.

The Sun's own orbit, Fairbridge found, has eight characteristic patterns, all determined by Jupiter's position relative to Saturn, with the other planets playing much lesser roles. Some of these eight have orderly orbits, smooth and near circular. During such orbits, solar activity is high and Earth heats up. Some of the eight orbits are chaotic, taking a loop-the-loop path. These orbits correspond to quiet times for the Sun and cool periods on Earth. Periodically the Sun embarks on a new cycle of orbits, which can be correlated with climate changes on Earth. There is more than one set of cycles giving rise to varying periodicities. If Fairbridge is right, the next cool period began in 1996, with the effects to be felt starting in 2010.

Temperatures on Earth are but one consequence of these periodic celestial movements. Others, Dr. Fairbridge argues, are seen everywhere on Earth: in the various and differing periodicities in rocks, glaciers, sand dunes, and the circulation of the ocean; geomagnetic records; the records of the isotopes of carbon, oxygen, and hydrogen in tree rings, ice cores, air, and water. These are the periodicities of climate change.

Dr. Fairbridge's best known periodicity, which he developed in the 1950s, hypothesized that sea levels had been rising for the last 16,000 years, during which time there were periodic rapid oscillations of rise and fall. The Fairbridge curve describing this period—so named in derision because it offended conventional wisdom—is now widely accepted. It demonstrates that, even within the past 1,000 years, sea levels have several times changed by up to two meters and—suddenly—each of these large changes occurred in fewer than 40 years. Precisely because his theory was so widely ridiculed at the time he put it forth, and is so widely accepted now, his ideas about global warming are taken seriously by those who know his work. As Richard Mackey wrote recently in the *Journal of Coastal Research*, "Rhodes Fairbridge was the first to document that the ocean levels rose and fell over long time scales. . . . On the basis of this work, Rhodes formulated the hypothesis that sea levels had been rising for the last 16,000 years and that the rise showed regular periodic oscillations of rise and fall over the period. This hypothesis, radical for its time and roundly rejected, is now acknowledged as a feature of the history of the planet."[34]

Dr. Fairbridge's broader climate-change claims—that celestial changes control Earth's temperatures—remain controversial, but less so than they were decades ago when his was a relatively lone voice. Dr. Fairbridge saw his Fairbridge curve theories vindicated, but he won't his celestial claims. This most remarkable individual died in 2006 at age ninety-two.

CHAPTER TWELVE

Better Safe Than Sorry?

William Gray, Cliff Ollier, Paul Reiter

When I launched my newspaper series, I naturally started by writing about deniers who contradicted headline-making claims. Most of those early columns—Landsea on hurricanes, Tol on the doomsday predictions of the Stern report, Wingham on the melting of Antarctica—naturally dealt with horror stories, because horror stories made headlines. Al Gore's book, in fact, is little more than a catalog of horrors.

But when I began profiling deniers who disputed not the horror stories but the actual physical case for human-caused catastrophic warming, a curious thing happened. I found myself further and further from the headlines and Al Gore's sensationalism. The media pays little or no attention to scientists such as Akasofu or Shaviv or the Danes or Abdussamatov. Wegman, himself, who ought to be famous, given his central place in one of the most important global warming arguments ever, remains an obscure figure except among his peers. The actual science—

even when it cuts in favor of the doomsayer case—attracts little attention.

I still occasionally do stories on "horror headline" issues. I profiled Dr. William Gray, from Colorado State University, who like Landsea is part of the hurricane story. For many years Dr. Gray headed up the university's Tropical Meteorology Project, which publishes yearly forecasts for tropical storms, named storms, typhoons, hurricanes, and intense hurricanes. This work has made him perhaps the world's foremost authority on the prediction of hurricanes, a science he pioneered and to which he has devoted more than 50 years of research. He and his colleagues have now reached a 95% accuracy rate in predicting the number of major storms and hurricanes that will occur in the next season. Insurance companies set their premiums, and government emergency-preparedness authorities set their budgets, on the basis of his pioneering work.

The essence of Gray's work has been to build more effective predictive models, based on deep empirical research. (He *likes* flying through hurricanes.) The experience has made him deeply skeptical about models loaded with fudge factors. He has described recent climate-change science as meaningless "mush," the product of simplistic computer models that crudely track a handful of factors and ignore the myriad others that influence the weather. This has earned him the usual pariah status including cutbacks in funding. In response, he has been forced to cut back on his research, contributing at least $100,000 from his own retirement account to keep his most important work going.

His work is a very direct rebuke to the catalog-of-horrors, anecdote-as-evidence approach. We may have had a few bad storm seasons, but there is no trend; the anecdotal "evidence" of 2004 and 2005 collapses in the face of the statistical record.

"The most reliable long-period hurricane records we have are the measurements of U.S. land-falling tropical cyclones since 1900 [Figure 1]. Although global mean ocean and Atlantic surface temperatures have increased by about 0.4°C between these two 50-year periods (1900–1949 compared with 1956–2005), the frequency of U.S. landfall numbers actually shows a slight downward trend for the later period. If we chose to make a similar comparison between U.S. landfall from the earlier 30-year period of 1900–1929 when global mean surface temperatures were estimated to be about 0.5°C colder than they were during the 30-year period from 1976–2005, we find exactly the same U.S. hurricane landfall numbers (54 to 54) and major hurricane landfall numbers (21 to 21)."[1]

No increase in named storms, hurricanes, or intense hurricanes making U.S. landfall during last 50 years

YEARS	Named Storms	Hurricanes	Intense Hurricanes (Cat 3-4-5)	Global Temperature Increase
1900-1949 (50 years)	189	101	39	+0.4°C
1956-2005 (50 years)	165	83	34	

Figure 1. U.S. landfalling tropical cyclones by intensity during two 50-year periods. *Adapted from* Philip J. Klotzbach and William M. Gray with special assistance from William Thorson, "Summary of 2007 Atlantic Tropical Cyclone Activity and Verification of Author's Seasonal and Monthly Forecasts," Department of Atmospheric Science, Colorado State University, November 27, 2007, p. 46, table 15. See: http://typhoon.atmos.colostate.edu/forecasts/2007/nov2007/nov2007.pdf.

One tip that anecdote is masquerading as evidence is the lack of a coherent physical theory. Professor Cliff Ollier applies that critique to headline stories like the June 2007 "Big Thaw" issue of *National Geographic*, with its awesome account of the sudden, rapid disappearance of the world's glaciers. Brilliantly illustrated with

THE CV OF A DENIER
Dr. William Gray

William Gray is professor emeritus of Atmospheric Science at Colorado State University, where he has worked since 1961. His many tropical field experiments were directed to the study of cumulus convection, condensation heating, evaporation cooling, sea-air energy-moisture exchange, and hurricane formation. He heads the university's Tropical Meteorology Project, which publishes yearly forecasts for tropical storms, named storms, typhoons, hurricanes, and intense hurricanes. He holds M.Sc. and Ph.D. degrees in meteorology and geophysical sciences from the University of Chicago.

the stunning photography for which the *Geographic* is famous, it prompted headlines around the world. Dramatized and magnified on nightly newscasts and blogs alike, it has become a staple in the popular imagination.

Unfortunately, the story as told by the *Geographic* is simply nonsense, claims Ollier. "Rapid melting of the Greenland and Antarctic ice sheets" in the manner described by the article and constantly repeated in the popular press, "is impossible"[2] he declares.

The scenarios raised in the *National Geographic* article rely on imaginary glaciers and ice sheets, not on the actual formations that exist in Greenland and Antarctica. These scenarios demonstrate a fundamental misunderstanding of the physics of glaciers and how they flow, says Ollier. For example, the models rely "on the concept of an ice sheet sliding down an inclined plane on a base lubricated by meltwater, which is itself increasing because of global warming,"[3] Prof. Ollier explains. "In reality, the Greenland

and Antarctic ice sheets occupy deep basins and cannot slide down a plane."[4]

Furthermore, ice cores, revealing complete records of depositions over periods of hundreds of thousands of years, show that the ice sheets have accumulated without melting even when temperatures have been warmer than now. "Ice sheets do not melt from the surface down, only at the edges,"[5] Prof. Ollier explains. The modelers' mechanism that has "meltwater lakes on the surface finding their way down through cracks in the ice and lubricating the bottom of the glacier is not compatible with accumulation of undisturbed snow layers."[6]

Ollier says the rate of ice flow now seen in the Polar Regions does not depend on the present climate but on the accumulation of ice that occurred in the distant past. The records "do not fit the model of surface melting, even infrequently. After three-quarters of a million years of documented continuous accumulation, how can we believe that right now the world's ice sheets are collapsing!"[7]

Of course the catalog-of-horrors approach to the global warming debate isn't just about chasing headlines. There is an animating idea behind it. Environmentalists call it the "Precautionary Principle."[8] One version of it goes as follows: "When an activity raises threats of harm to human health or the environment, precautionary measures should be taken even if some cause-and-effect relationships are not fully established scientifically."[9] Or as your mother would have said, "better safe than sorry."

The problem with the better-safe-than-sorry approach to headline horrors is that the headlines so often finger the wrong culprit. That's very dangerous because the real culprits keep getting away with it—especially if the culprits are some of the world's worst governments. Much of my work as an environmentalist has been trying to hold governments to account for wickedly destructive development projects—like mega-hydropower projects in

China displacing millions of poor people from their homes. Yet these governments continue to get away with defending such projects as "green."

Headline horrors make great scapegoats. There's no more egregious or vicious example than governments using global warming to cover up their own failures to prevent the resurgence of malaria and other mosquito-borne diseases.

Once on the verge of eradication, except in sub-Saharan Africa, today malaria "is again common in many parts of Central America, the northern half of South America, much of tropical and subtropical Asia, some Mediterranean countries and many of the republics once part of the Union of Soviet Socialist Republics."[10] Travelers from these regions can spread the disease when local mosquitoes pick it up and become transmitters. "Indigenous transmission associated with imported cases has recently been reported in Kazakhstan, Kyrgyzstan, Turkmenistan, Uzbekistan, Bulgaria, the Republic of Moldova, Romania, Italy, and Corsica, and the malaria-free status of Europe may be in jeopardy."[11]

This is a very real and thoroughly documented public health problem. But why is it happening now?

THE CV OF A DENIER
Dr. Cliff Ollier

Dr. Cliff Ollier, a geologist, geomorphologist, and soil scientist is emeritus professor and honorary research fellow, University of Western Australia. He has authored or coauthored more than 500 publications, mostly in world-class journals, and ten books, including the widely acclaimed *Tectonics and Landforms* and *The Origin of Mountains*. He received his doctorate from Bristol University.

Al Gore says he knows. He wrote in his book *An Inconvenient Truth*: "In general, the relationship between the human species and the microbial world of germs and viruses is less threatening when there are colder winters, colder nights, more stability in climate patterns, and fewer disruptions. . . . To cite one important example of this phenomenon, mosquitoes are profoundly affected by global warming. There are cities that were originally located above the mosquito line, which used to mark the altitude above which mosquitoes would not venture. Nairobi, Kenya and Harare, Zimbabwe are two such cities. Now with global warming, the mosquitoes are climbing to higher altitudes."[12] How does Gore know this? The IPCC told him so way back in 1995 when it issued its *Second Assessment Report*. In one of the chapters on human population health, the IPCC created the scare—repeated by scientists with a popular following—that global warming could lead to 80-million additional cases of malaria per year worldwide.[13]

There's just one problem. This claim reflects such "glaring . . . ignorance"[14] of the subject that it would be laughable if it were not so dangerous. The resurgence of malaria and other mosquito-borne diseases is very real. But, say the real experts, climate change is at most a trivial factor. Following Gore's advice rather than addressing the real causes could mean grave illness or death for millions.

Professor Paul Reiter heads the Insects and Infectious Diseases Unit at the Pasteur Institute, famed for its founding by Louis Pasteur in 1887 and the eight Nobel Prizes that its later scientists received. Prior to joining the Pasteur Institute, Prof. Reiter directed the Entomology Section at the Dengue Branch of the Centers for Disease Control, the path-breaking U.S. government agency. Prof. Reiter is also known for his work as an officer of the Harvard School of Public Health, his membership on the World Health Organization's Expert Advisory Committee on Vector Biology and Control, and, among administrative positions, his role as lead author of the Health Section of the U.S. National Assessment of the Potential Consequences of Climate Variability and Change.

"I know of no major scientist with any long record in this field who agrees with the pronouncements of the alarmists at the IPCC,"[15] states Prof. Reiter, whose history in his research field spans three decades and five continents, and who is well familiar with the scope of work occurring in the mosquito-borne research community. "On the contrary, all of us who work in the field are repeatedly stunned by the IPCC pronouncements. We protest, but are rarely quoted, and if so, usually as a codicil to the scary stuff."[16]

Reiter says that using climate models to predict the spread of mosquito-borne diseases, as the IPCC does, reflects a dangerous ignorance. "[T]he histories of three such diseases—malaria, yellow fever, and dengue—reveal that climate has rarely been the principal determinant of their prevalence or range." Climate factors, "particularly rainfall, are sometimes—but by no means always—relevant."[17] The "principal determinants are politics, economics, and human activities."[18] In brief, sound public health policies are the key to these diseases. When sound policies are followed, the mosquitoes and the diseases disappear. When they are abandoned, the diseases return.

"The current increase in malaria is alarming,"[19] he fervently declares. But, he says, "the principal factors involved are deforestation, new agricultural practices, population increase, urbanization, poverty, civil conflict, war, AIDS, resistance to anti-malarials, and resistance to insecticides, not climate. In my opinion, we should give priority to a creative and organized effort to stem the burgeoning tragedy of uncontrolled malaria, rather than worrying about the weather."[20]

The doomsayer case is based on the proposition that mosquito-borne diseases have been heretofore limited by two factors: latitude and altitude. Malaria, they say, in the past has not been a problem in the temperate zones. And in warmer latitudes it has not afflicted cities located at lofty altitudes, also allegedly too cold and too dry to support mosquitoes. The spread of the diseases to higher latitudes and altitudes is caused by global warming.

These claims, says Reiter, reflect an astonishing ignorance of disease history.

"Until the second half of the 20th century," he writes, "malaria was endemic and widespread in many temperate regions, with major epidemics as far north as the Arctic Circle. From 1564 to the 1730s—the coldest period of the Little Ice Age—malaria was an important cause of illness and death in several parts of England."[21] The disease began to decline only in the 19th century, "when the present warming trend was well under way."[22]

Malaria was common throughout much of the United States until the 20th century, "and it remained endemic until the 1950s. . . . Widespread epidemics of dengue were also common, and continued until the 1940s. In Europe, malaria was probably present in neolithic times. In ancient Greece, Hippocrates clearly distinguished between the symptoms of vivax and falciparum malaria."[23] Historically nearly all Europe suffered. As for other mosquito-borne diseases, "yellow fever also killed tens of thousands in many European countries until the end of the 19th century, and a devastating epidemic of dengue, with an estimated 1 million cases and 1,000 deaths, occurred in Greece in 1927–28."[24] The disease was largely eliminated in Europe as in the United States by "changes in agriculture and lifestyle that affected the abundance of mosquitoes, their contact with people, and the availability of anti-malarial drugs."[25]

And when the disease has resurged in the developed world, the cause clearly has not been climate. Reiter points out that "the most catastrophic epidemic on record anywhere in the world occurred in the Soviet Union in the 1920s."[26] At the time, the country was suffering a massive breakdown in civic and public services in the wake of invasion, revolution, and civil war. The epidemic peaked at "13 million cases per year, and 600,000 deaths. Transmission was high in many parts of Siberia, and there were 30,000 cases and 10,000 deaths in Archangel, close to the Arctic Circle."[27]

Well what about "altitude"? Is the "mosquito line" rising as Gore and the IPCC tell us? More nonsense says Reiter. Perhaps chastened by the news that the greatest malaria epidemic ever recorded extended into northern Siberia, recently the doomsayers have shifted their emphasis "to 'highland malaria,' particularly in East Africa."[28] Ignoring serious research, "a flurry of articles by non-specialists . . . claim a recent increase in the altitude of malaria transmission attributable to warming, and quote models that 'predict' further increase in the next 50 years."[29] But, says Reiter, the records show cases from 1880 to 1945 that "were 500–1,500 m higher than in the areas that are quoted as examples."[30]

In fact, he points out, "highland malaria was widespread throughout the world until the era of DDT and cheap malaria prophylaxis."[31] Figure 2 "shows the maximum altitude of autochthonous cases in 11 countries in the early half of [the 20th] century. Transmission occurred to 2,600 m in Kenya, and 2,450 m in Ethiopia. In the Himalayas, the disease was present to 2,500 m in India and 1,830 m in China. In the Andes, epidemics were recorded to 2,180 m in Argentina and 2,600 m in Bolivia. In the latter country, cases actually occurred to 2,773 m, transmitted by mosquitoes breeding at 35°C in thermal springs."[32]

The doomsayers have attributed recent malaria epidemics in the highlands of Madagascar to "global warming, although they occurred well below the maximum altitude for transmission"[33] and followed a breakdown in mosquito-control efforts. "Moreover, similar epidemics had taken place in the same areas in 1878 and 1895, and local records show no great change in temperature."[34]

Malaria was common at high altitudes in the first half of the 20th century

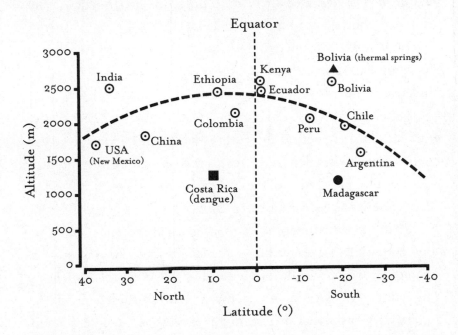

Figure 2. Latitude and upper altitude limits of malaria transmission in 11 countries before 1945. The curve shows the approximate relation between maximum latitude and altitude. Note that the 1987 malaria epidemic in Madagascar (solid circle) and the 1993 dengue epidemic in Costa Rica (solid square) are both well below the curve. Also local conditions may support outbreaks at anomalously high altitudes, as in the hot springs in Bolivia. Reiter's figure based on Hackett, 1945. *Adapted from* Paul Reiter, Correspondence: "Global-Warming and Vector-borne Disease In Temperate Regions and at High Altitude," Correspondence and author's reply, *The Lancet* 351 (March 14, 1998): 839-840. See http://www.sepp.org/Archive/controv/controversies/lancetltrs.html.

One problem appears to be that the doomsayers either do not read the research in detail, or if they do, they distort it. For instance, "repeated claims that the disease has ascended to new altitudes in Colombia consistently cite a publication by Nelson and Suarez et al. [M. N. Nelson and M. F. Suarez et al., "The Distribution of Aedes aegypti at High Elevations in Colombia," World

Health Organization WHO/VBC/83.872, 1983] but ignore its content, for although the vector was present to 2,200 m, the investigators clearly stated there were no cases at high altitude, and none have been reported since that study."[35]

How can this happen? How could the UN have issued such dangerously amateurish advice on such a crucial issue, one that threatens hundreds of millions of people?

For Reiter the answer is simple. First of all, those 2,500 scientists the UN trumpets as the authors and reviewers of the IPCC reports aren't what they seem to be, and also aren't very good, at least not in Reiter's area of expertise. As Prof. Reiter testified to a UK parliamentary committee in 2005, "The paucity of information"[36] in the IPCC reports "was hardly surprising: not one of the lead authors had ever written a research paper on the subject! Moreover, two of the authors, both physicians, had spent their entire careers as environmental activists. One of these activists has published 'professional' articles as an 'expert' on 32 different subjects, ranging from mercury poisoning to land mines, globalization to allergies, and West Nile virus to AIDS."[37] According to Reiter, the contributing authors included exactly one "professional entomologist, and a person who had written an obscure article on dengue and El Niño, but whose principal interest was the effectiveness of motorcycle crash helmets (plus one paper on the health effects of cell phones)."[38]

Moreover, the IPCC has twisted the peer-review process—essential to ensuring rigorous science. In professional science, "articles submitted to journals for publication are sent to persons in the same field for 'Peer Review.' Up to five such peers are asked to write a critique, and to recommend acceptance, modification, or rejection."[39] The names of peer reviewers are kept confidential to encourage independent criticism, free of recrimination, while the deliberations of the authors being critiqued are made public.

THE CV OF A DENIER
Paul Reiter

Paul Reiter, professor at the Pasteur Institute in Paris, is chief of its Insects and Infectious Diseases Unit and a specialist in the natural history and biology of mosquitoes, the epidemiology of the diseases they transmit, and strategies for their control. He was chairman of the American Committee of Medical Entomology of the American Society for Tropical Medicine and Hygiene and of several committees of other professional societies. He has worked for the World Health Organization, the Pan American Health Organization, and other agencies in investigations of outbreaks of mosquito-borne diseases, as well as of AIDS and Ebola haemorrhagic fever and onchocerciasis. He was also a contributory author of the IPCC *Third Assessment Report*. He has been chairman of the American Committee of Medical Entomology of the American Society for Tropical Medicine and Hygiene and of several committees of other professional societies. He received his Ph.D. in Medical Entomology from the University of Sussex in 1978.

"The IPCC turns this on its head,"[40] Prof. Reiter explains. "The peer reviewers have to give their names to the authors, but the deliberations of the authors are strictly confidential."[41] In effect, the science is spun, disagreements purged, and results predetermined.

The real problem, Prof. Reiter believes, is that "[t]he Intergovernmental Panel is precisely that—it is a panel among governments."[42] Prof. Reiter wrote the IPCC with a series of detailed questions about its decision-making process. It replied: "The brief answer to your question below is 'governments.' It is the governments of the world who make up the IPCC, define its remit, and direction. The way in which this is done is defined in the IPCC Principles and Procedures, which have been agreed by governments."[43]

But in the case of malaria, for example, governments are the problem. Reiter notes that each of the possible causes of the "rapid recrudescence" of mosquito-borne diseases—including population growth, "irrigation and other agricultural activities, ecologic change, movement of people, urbanization, deterioration of public health services, resistance to insecticides and anti-malarial drugs, deterioration of vector control operations, and disruptions from war, civil strife, and natural disasters."[44]—is at root a political problem, a failure of governments. Climate change is a terribly convenient scapegoat for governments, which control the IPCC.

CHAPTER THIRTEEN

Some Inconvenient Persons

Roger Revelle, Claude Allegre, Reid Bryson, David Bellamy

O ver the course of writing this book, I consistently found that the deniers were at least as credible and often more credible than the doomsayers with whom they had locked horns. In many cases it was the IPCC itself that testified to their credibility by including them in the process.

At that level then, our mission is accomplished. Knowing what we know now it is not possible to believe that the science is settled or that there is a scientific "consensus" for the doomsayer view of global warming.

But this raises another issue. What are we to think now of the people who so relentlessly claim "the science is settled," who have tried so hard to persuade us that the only dissent is from the moral equivalent of flat-earthers or holocaust deniers? How are we to avoid the conclusion that the doomsayers will say anything—and slander anyone—to defend their thesis? What about their credibility?

In the history of the global warming movement, no scientist is more revered than Roger Revelle of the Scripps Institution of Oceanography, Harvard University, and University of California San Diego. He was the coauthor of the seminal 1957 paper that demonstrated that fossil fuels had increased carbon dioxide levels in the air. Under his leadership, the President's Science Advisory Committee Panel on Environmental Pollution in 1965 published the first authoritative U.S. government report in which carbon dioxide from fossil fuels was officially recognized as a potential global problem. He was the author of the influential 1982 *Scientific American* article that elevated global warming to the public agenda. For being "the grandfather of the greenhouse effect,"[1] as he put it, he was awarded the National Medal of Science by the first president Bush.

Roger Revelle's most consequential act, however, may have come in his role as a teacher. During the 1960s at Harvard, Dr. Revelle inspired a young student named Al Gore. Dr. Revelle would change Gore's life, giving Gore and his fellow students advance notice of the fruits of his cutting-edge research on climate change.

"It felt like such a privilege to be able to hear about the readouts from some of those measurements in a group of no more than a dozen undergraduates,"[2] Gore later explained. "Here was this teacher presenting something not years old but fresh out of the lab, with profound implications for our future!"[3]

Calling him "a wonderful, visionary professor"[4] who was "one of the first people in the academic community to sound the alarm on global warming,"[5] Gore thought of Dr. Revelle as his mentor and referred to him frequently, relaying his experiences as a student in his book *Earth in the Balance*, published in 1992. Gore's warmth for Dr. Revelle cooled, however, when it became clear that he had misunderstood his former professor. Although Dr. Revelle recognized potential harm from global warming, he also saw potential benefits and was by no means alarmed, as indicated in this 1984 interview in *Omni* magazine:

> **Omni:** A problem that has occupied your attention for many years is the increasing levels of CO2 in the atmosphere, which could cause the Earth's climate to become warmer. Is this actually happening?
>
> **Revelle:** I estimate that the total increase [in CO2] over the past hundred years has been about 21%. But whether the increase will lead to a significant rise in global temperature, we can't absolutely say.
>
> **Omni:** What will the warming of the Earth mean to us?
>
> **Revelle:** There may be lots of effects. Increased CO2 in the air acts like a fertilizer for plants . . . you get more plant growth. Increasing CO2 levels also affect water transpiration, causing plants to close their pores and sweat less. That means plants will be able to grow in drier climates.
>
> **Omni:** Does the increase in CO2 have anything to do with people saying the weather is getting worse?
>
> **Revelle:** People are always saying the weather's getting worse. Actually, the CO2 increase is predicted to temper weather extremes. . . . [6]

A few years later, in a July 14, 1988, letter to Congressman Jim Bates, Revelle was still urging caution. "Most scientists familiar with the subject are not yet willing to bet that the climate this year is the result of 'greenhouse warming.' As you very well know, climate is highly variable from year to year, and the causes of these variations are not at all well understood. My own personal belief is that we should wait another ten or twenty years to really be convinced that the greenhouse effect is going to be important for human beings, in both positive and negative ways."[7]

Just a few days later on July 18, Revelle again warns against leaping to conclusions, this time in a letter to Senator Tim Wirth: "We should be careful not to arouse too much alarm until the rate and amount of warming becomes clearer. It is not yet obvious that this summer's hot weather and drought are the result of a global climatic change or simply an example of the uncertainties of climate variability. My own feeling is that we had better wait another ten years before making confident predictions."[8]

Then in 1991, Dr. Revelle collaborated on an article for the *Cosmos: A Journal of Emerging Issues* with two illustrious colleagues, Chauncey Starr, founding director of the Electric Power Research Institute, and Fred Singer, the first director of the U.S. Weather Satellite. Entitled "What To Do About Greenhouse Warming: Look Before You Leap,"[9] the article argued that decades of research could be required for the consequences of increased carbon dioxide to be understood and laid out the harm that could come of acting recklessly: "Drastic, precipitous, and especially, unilateral steps to delay the putative greenhouse impacts can cost jobs and prosperity and increase the human costs of global poverty, without being effective. Stringent controls enacted now would be economically devastating, particularly for developing countries for whom reduced energy consumption would mean slower rates of economic growth without being able to delay greatly the growth of greenhouse gases in the atmosphere. . . . 'Look before you leap' may still be good advice."[10]

Three months after the *Cosmos* article appeared, Dr. Revelle died of a heart attack. One year later, with Al Gore running for vice president in the 1992 presidential election, the inconsistency between Gore's pronouncements—he claimed that the science was settled then too—and those of his mentor, became national news, thanks to an article in the July 6, 1992, issue of the *New Republic*

THE CV OF A DENIER
Roger Revelle

Roger Revelle was professor of oceanography at Scripps Institution of Oceanography and became its director from 1950 to 1964. After his successful efforts to create the University of California San Diego, he went to Harvard University, where he was professor of Population Policy and director of the Center for Population Studies until 1976. He was also founding chairman of the first Committee on Climate Change and the Ocean under the Scientific Committee on Ocean Research and the International Oceanic Commission. Dr. Revelle received a Ph.D. in oceanography from UC Berkeley in 1936.

by Gregg Easterbrook. The subject even came up in the vice-presidential debates that year. Gore counterattacked, leading to claims that Dr. Revelle had become senile before his death, that Dr. Singer had duped Dr. Revelle into coauthoring the article, and that Dr. Singer had listed Dr. Revelle as a coauthor over his objections.

The vehicle for these claims was Dr. Justin Lancaster, a Harvard scientist who had known Dr. Revelle, and was an advisory editor for an anthology that planned to reprint the *Cosmos* article. According to his own deposition in a later libel suit filed by Singer, after the *Cosmos* article became a public issue, Gore called Lancaster and inquired about Revelle's mental capacity in the year before his death and whether Revelle really believed what was attributed to him in the *Cosmos* article. At the time, Lancaster drafted (but apparently never sent) a letter to Gore affirming that Revelle was mentally fit and "not casual about his integrity."[11] In the draft letter, Lancaster wrote that Revelle "felt it was honest to admit the uncertainties [about greenhouse warming], including the idea that our ignorance could be hiding benefits as well as catastrophes."[12] Later, however, senior-Gore-staff people began discussing with Lancaster the possibility of getting Revelle's name removed from reprints of the article. Lancaster also apparently submitted to Gore's staff numerous drafts of a proposed article for the *New Republic* responding to the Easterbrook article.[13]

In July 1992, Lancaster called Dr. Singer, one of the two surviving coauthors of the article. According to Singer, Lancaster first requested, and then demanded, that Revelle's name be removed from the article in the anthology and from all future reprints. Singer refused, as did the publishers of the anthology. Subsequently, in a memorial symposium for Revelle, Lancaster publicly charged that Revelle's name had been placed on the article over his objections. He later added the charge that Revelle's mental capacities were failing at the time, implying Singer had taken advantage.

Eventually, Singer sued. In discovery and during the course of the trial, several crucial facts came out:

- Singer had in his possession the marked-up galleys that he and Revelle had reviewed together to make final changes in the article, including extensive notes in Revelle's own hand.

- Lancaster had been aware of the article long before he made any objection to it, which he did only after it became a public problem for Gore.

- Lancaster's extensive contacts with Gore and his staff were documented.

- Revelle's secretary provided Revelle's schedule at the time the article was being written, edited, and approved, showing he was speaking, traveling, and attending conferences throughout the period, and apparently in full possession of his faculties. However, in her deposition she also testified that Revelle did not seem eager to do the *Cosmos* article and that Singer had to do a fair bit of nagging to get Revelle to finish the work.[14]

Ultimately, Lancaster settled the case, offering a complete retraction, which read in part, "I retract as being unwarranted any and all statements, oral or written, I have made which state or imply that Professor Revelle was not a true and voluntary coauthor of the *Cosmos* article, or which in any other way impugn or malign the conduct or motives of Professor Singer with regard to the *Cosmos* article (including but not limited to its drafting, editing, publication, republication, and circulation). I agree not to make any such statements in future. I fully and unequivocally retract and disclaim those statements and their implications about the conduct, character, and ethics of Professor Singer, and I apologize to Professor Singer for the pain my conduct has caused him and for any damage that I may have caused to his reputation."[15]

Subsequently, Gore went after Singer, calling Ted Koppel and demanding he investigate Singer's sources of funding. Koppel did look into the case and responded by very publicly rapping Gore's knuckles: "There is some irony in the fact that Vice President Gore—one of the most scientifically literate men to sit in the White House in this century—[is] resorting to political means to achieve what should ultimately be resolved on a purely scientific basis."[16]

Quite recently, Lancaster retracted his retraction, claiming he had only issued the retraction in the first place because of the financial strain of the lawsuit. His retraction of the retraction, however, though strident rhetorically, is somewhat guarded as to assertions of fact, dwelling on the point that Singer wrote the original draft.[17] This, Singer has never disputed. Nor is it uncommon practice in the scientific world for one (often junior) author to take the initiative and recruit other authors to join in a work, because they have relevant specialized knowledge and would add credibility to the article. It was always quite clear from Singer that both Revelle and the other coauthor, Dr. Chauncey Starr, winner of the National Technology Medal among other honors, had come in on that basis.

A 1992 letter from Revelle's daughter and other family members criticized columnist George Will for implying that Revelle had come to doubt the existence of global warming. However, the letter, which we quote at length below, clearly affirms that Revelle did indeed urge a cautious response to global warming and certainly did not regard the science as settled. Also the letter cites the *Cosmos* article in support of the family's account of Revelle's views, implicitly affirming that the article

'Global Warming: What My Father Really Said'

Carolyn Revelle Hufbauer,
The Washington Post, September 13, 1992

Contrary to George Will's "Al Gore's Green Guilt" {op-ed, Sept. 3} Roger Revelle—our father and the "father" of the greenhouse effect—remained deeply concerned about global warming until his death in July 1991. That same year he wrote: "The scientific base for a greenhouse warming is too uncertain to justify drastic action at this time." Will and other critics of Sen. Al Gore have seized these words to suggest that Revelle, who was also Gore's professor and mentor, renounced his belief in global warming.

Nothing could be [further] from the truth.

When Revelle inveighed against "drastic" action, he was

accurately represented those views. Revelle's daughter points out that Revelle urged a number of measures to conserve energy and reduce our use of fossil fuels. So do I. Nothing in the family letter makes him an unlikely coauthor of the *Cosmos* article—or a likely co-doomsayer with Al Gore.

Who is right? We know that in the *Omni* interview years earlier, and in his 1988 correspondence with Representative Bates and Senator Wirth, Revelle expressed views very like what he is alleged to have written in the *Cosmos* article. These documents are damning to the idea that Revelle would approve of his most famous student's doomsaying. And we know that rather than welcome the thoughtful cautions of his former professor, which after all amounted to little more than "let's be sure," Gore's response was to look for evidence that the old man—his revered mentor—was senile.

using that adjective in its literal sense—measures that would cost trillions of dollars. Up until his death, he thought that extreme measures were premature. But he continued to recommend immediate prudent steps to mitigate and delay climatic warming. Some of those steps go well beyond anything Gore or other national politicians have yet to advocate. . . .

So in recent speeches and writings, he recommended several kinds of action, including. . . .

Conserve energy. Revelle advocated conserving energy by using the price mechanism (the polluter pays principle)—for example, by increasing the tax on gasoline (*Cosmos*, 1991). In private, he often spoke of a $1.00 a gallon tax as eminently reasonable, not "drastic." Who was the last national politician to advocate a $1.00 gasoline tax?

Roger Revelle proposed a range of approaches to address global warming. Inaction was not one of them. He agreed with the adage "look before you leap," but he never said "sit on your hands. . . ."

Revelle is the most famous case of a one-time environmental hero being tossed over the side when he became inconvenient, but hardly the only one.

Claude Allegre, one of France's leading socialists and among her most celebrated scientists, was among the first to sound the alarm about the dangers of global warming. "By burning fossil fuels, man increased the concentration of carbon dioxide in the atmosphere which, for example, has raised the global mean temperature by half a degree in the last century,"[18] Dr. Allegre, a geochemist, wrote 20 years ago. Fifteen years ago, Dr. Allegre was among the 1,500 prominent scientists who signed "World Scientists' Warning to Humanity," a highly publicized letter stressing that global warming's "potential risks are very great"[19] and demanding a new caring ethic that recognizes the globe's fragility in order to stave off "spirals of environmental decline, poverty, and unrest, leading to social, economic, and environmental collapse."[20]

That was back in the days before the world had spent billions studying the problem. Then the evidence began to come in, and Dr. Allegre . . . changed his mind. To his surprise, the many climate models and studies failed dismally in establishing a manmade cause of catastrophic global warming. On his reading, the evidence indicated most of the warming comes from natural phenomena. Dr. Allegre now sees global warming as overhyped and an environmental concern of second rank.

His public break with what he now sees as environmental cant on climate change came in September 2006 in an article entitled "The Snows of Mount Kilimanjaro" in *L' Express*, the French weekly. His article cited evidence that Antarctica is gaining ice and that Kilimanjaro's retreating snowcaps, among other global warming concerns, come from natural causes. "The cause of this climate change is unknown,"[21] he states matter-of-factly.

Allegre is an exalted member of France's political establishment, a friend of former Socialist prime minister Lionel Jospin, and from 1997 to 2000, minister of education, research, and technology. Dr. Allegre has the highest environmental credentials. The author of early environmental books, he fought successful battles to protect the ozone layer from CFCs and public health from lead pollution. He is, above all, a scientist of the first order, the architect of isotope geodynamics, which showed that the atmosphere was primarily formed early in the history of the Earth, and the geochemical modeler of the early solar system. Because of his path-breaking cosmo-chemical research, NASA asked Dr. Allegre to participate in the Apollo Lunar Program, where he helped determine the age of the moon. Dr. Allegre is perhaps best known for his research on the structural and geochemical evolution of the Earth's crust and the creation of its mountains: "The Snows of Mount Kilimanjaro" are a rather personal matter to him.

Allegre's break with scientific dogma over global warming came at some personal cost: colleagues in both the governmental and environmental spheres were aghast that he could publicly question the science behind climate change. But what does it say about the credibility of such doomsayer colleagues, that they can dismiss a man of such impeccable credentials, a man of their own political persuasion who has fought the great environmental battles of our time, merely for voicing doubt?

Sometimes even the most globally eminent scholars find themselves conveniently ignored once they dissent on global warming. Reid Bryson has been called the "father of scientific climatology." He holds the title "the world's most cited climatologist," according to an analysis in the journal of the Institute of British Geographers. He's also the fifth most cited physical geographer in the world, and the eleventh most cited among all geographers. He has written some 230 articles and five books, including in such fields as geology, limnology, meteorology, and archeology.

THE CV OF A DENIER
Claude Allegre

Claude Allegre received a Ph.D. in physics in 1962 from the University of Paris. He became the director of the Geochemistry and Cosmochemistry Program at the French National Scientific Research Center in 1967, and in 1971, he was appointed director of the University of Paris's Department of Earth Sciences. In 1976, he became director of the Paris Institut de Physique du Globe. He is an author of more than 100 scientific articles, many of them seminal studies on the evolution of the Earth using isotopic evidence, and 11 books. He is a member of the U.S. National Academy of Sciences and the French Academy of Science.

But he is not just an eminent scientist. He has been for decades a hero to the environmental movement. He has twice seen his papers in *Environmental Conservation* awarded prizes for being "best paper of the year." He's a member of the United Nations Global 500 Roll of Honor, created to recognize "outstanding achievements in the protection and improvement of the environment." Those of us with long histories in the movement remember him as an inspirational figure as far back as the 1970s, challenging the wasteful ways of our consumer society, and warning of a dire need for lifestyle changes. *Mother Earth News*, a bible of the environmental movement, in the preamble to an extensive 1976 interview, described him as "an environmentalist in the broadest sense and his thoughts on the planet, its human population, and that population's activities range as widely and carry all the force of such acknowledged environmental spokesmen as Barry Commoner, Paul Ehrlich, and Dave Brower."[22]

Bryson's verdict on manmade global warming: "[It] is a theory for which there is no credible proof."[23] Certainly, he argues, humans affect the climate in ways that both warm and cool the atmosphere. His complaint was, and remains, that global warming alarmists cherry pick the evidence to focus on warming influences of human behavior, while not taking sufficient account of cooling influences. He has been delivering that message at least since 1976.[24]

But he is very old now, and so, though he is still professionally active, the doomsayers simply don't mention him at all. These days he argues that the global warming movement distorted science to pursue other agendas. "There is very little truth to what is being said and an awful lot of religion,"[25] he has decided. "It's almost a religion where you have to believe in anthropogenic global warming or else you are nuts."[26] And as for the often-claimed scientific consensus on climate change, he doubts it: "I know of no vote having been taken, and know that if such a vote were taken of those who are most vocal about the matter, it would include a significant fraction of people who do not know enough about climate to have a significant opinion."[27]

And as for the biggest believer, Al Gore, and his movie, *An Inconvenient Truth*: "Don't make me throw up," Bryson exclaims. "It is not science. It is not true."[28]

Of this group of inconvenient dissenters, none was more eye-opening than Professor David Bellamy, who is Great Britain's best-known environmentalist, and has been for most of the last four decades. He has written and presented some 400 television programs on environmental issues, written 45 books, and published more than 80 scientific papers, in addition to holding down teaching posts in botany at two universities. He has founded or been president of such prominent national environmental organizations as The Conservation Foundation, The

Royal Society of Wildlife Trusts, Population Concern, Plantlife International, British Naturalists' Association, and Galapagos Conservation Trust.

Among his many honors have been the United Nations Environmental Program Global 500 Award, the Duke of Edinburgh's Award for Underwater Research, and the Order of the British Empire. No mere academic, this legendary figure has a long and distinguished record as a fighter for green causes, starting with the 1967 Torrey Canyon supertanker disaster off the coast of England. He has led high-profile protests against needless road building and the loss of moors and has been jailed for blockading the construction of a hydro dam that would have destroyed a Tasmanian rainforest.

He was a hero to environmentalists around the world, that is, until July 9, 2004. That was the day a full-page article by him appeared in *London's Daily Mail*, summing up his views on global warming alarmism with the headline, "What a load of poppycock!"[29] In Prof. Bellamy's characteristic no-holds-barred style he declared, "Whatever the experts say about the howling gales, thunder, and lightning we've had over the past two days, of one thing we can be certain. Someone, somewhere—and there is every chance it will be a politician or an environmentalist—will blame the weather on global warming."[30] But, he proclaimed, "they will be 100% wrong. Global warming—at least the modern nightmare version—is a myth."[31]

Challenging the very premise of the doomsayer case, he pummeled the notion of CO_2 as a "dreaded killer greenhouse gas" when it "is, in fact, the most important airborne fertilizer in the world, and without it there would be no green plants at all."[32] Even doubling the amount of carbon dioxide in the atmosphere "would produce a rise in plant productivity. Call me a biased old

THE CV OF A DENIER
Reid Bryson

Reid Bryson joined the faculty of the University of Wisconsin-Madison in 1946, and in 1948, became the founding chairman of its Department of Meteorology. In 2007, he became emeritus professor of the university's Department of Oceanic and Atmospheric Sciences. Dr. Bryson's research established new paths through diverse fields, among them the Indian monsoon, airstreams, and the reconstruction of past climates. He is a fellow of the American Association for the Advancement of Science and of the American Meteorological Society. He received his Ph.D. in meteorology at the University of Chicago.

plant lover, but that doesn't sound like much of a killer gas to me. Hooray for global warming is what I say, and so do a lot of my fellow scientists."[33]

Prof. Bellamy, a passionate socialist as well as environmentalist, went on to argue against alarmist government action on global warming, fearing that billions or even trillions of dollars could be diverted "on a problem that doesn't exist—money that could be used in umpteen better ways: fighting world hunger, providing clean water, developing alternative energy sources, improving our environment, creating jobs."[34]

The reaction?

Plantlife International, the UK's leading charity dedicated to protecting wild plants, announced it "would be wrong to ask him to continue"[35] as president, a post which Prof. Bellamy had held for 15 years.

The Royal Society of Wildlife Trusts, which manages 2,500

THE CV OF A DENIER
David Bellamy

Professor David Bellamy, a botanist, is special professor of Geography at University of Nottingham and honorary professor of the University of Central Queensland.

His most recent paper, "Climate Stability: An Inconvenient Proof," published in the refereed civil engineering journal of the Institution of Civil Engineers in May 2007, demonstrates that, in the unlikely event the widely prophesied doubling of carbon dioxide levels from natural, pre-industrial levels occurs, the warming would amount to less than 1°C of global warming. Prof. Bellamy received his doctorate from the University of London.

nature reserves across the UK, likewise announced it would not renew his presidency.

Former colleagues in the environmental movement now suggested he had become mentally incompetent, or was in the pay of the oil industry.

And they have kept it up. In 2007, the Carbon Trust described him as a "[l]oony IPCC debunker," "climate change denying shill," and "the very sad and deluded David Bellamy."[36]

In the 1960s, before the era of environmental activism, Bellamy's was a lonely voice decrying an establishment insensitive to environmental dangers derided by the conventional wisdom of the times. As environmental awareness grew, he became an exemplar of the establishment. Now his is a lonely voice once more, as a new establishment once again concludes that dissent is grounds for dismissal.

CHAPTER FOURTEEN

The Carbon Catastrophe

This book must now end because my deniers are without end. When I started my "Denier" series in my newspaper column, I hoped it might run to three or even six installments—I had no expectation of finding more than a few credible scientists. The more I looked, the more I found, and as scientists around the world learned that someone was seeking them out, the more found me. I now have a backlog of some 100 deniers to write about, a backlog that grows weekly. I suspect most will never make it into my newspaper series, let alone into this book, which already excludes many fine scientists who have been part of my series, men at the highest levels of their disciplines, such as Michael Griffin, the head of NASA, Hans von Storch, perhaps Germany's best known climate scientist, and William Nordhaus, perhaps the most respected climate-change economist in the world.

Have my deniers convinced me that global warming is all a hoax? They have not, despite my admiration for the courage and integrity that so many of them display, and despite their impeccable credentials, credentials, I might add, that are often far more impressive than those of some of the gurus propounding climate-change catastrophes. Michael Mann, you might recall, was fresh out of school when he attained stardom for his hockey stick theory, and those who pushed the notion that global warming was responsible for Katrina and other hurricanes had paltry credentials in their field at best.

My deniers certainly demonstrate that the climate-change doomsayers should not have the last word, but they also demonstrate that they, themselves, can't have the last word either. After all, most of my deniers disagree with each other as well as with the doomsayers. They can't all be right. They could all be wrong. Just as the doomsayers, the great majority of whom I believe to be entirely sincere and highly qualified, could all be wrong.

How can I explain my respect for scientists on both sides of the IPCC divide? I will explain by analogy to the world of economics.

For the last ten years, in addition to my job at Energy Probe, I have edited the op-ed submissions of many impressive economists to the *National Post*, among them Nobel Laureates such as Milton Friedman and Robert Mundell. When these economists engaged in debate in the pages of the *Post* on the economic controversies of the day, they rarely agreed. In some cases, half a dozen economists would passionately tackle some issue or other and present half a dozen different opinions of what was going on. They could not agree on whether the stock market was going up or down, or the economy was headed into a recession, or whether the dollar was overvalued.

We all know enough to take the forecasts of economists with a grain of salt. It's not that we think they're pulling a hoax. But we know from long experience that economics, and especially economic models, encounter severe "limits to predictability," as Henk

Tennekes might say, no matter how many super-computers they use. Why should we have any more confidence in climate science? After all, economic models have far fewer variables to contend with than climate models. And the data that gets plugged into economic models is far less controversial than much of the data used by climate scientists. Economists for the most part agree on GDP and employment data and that sort of thing. Yet economists with their models are hopelessly inadequate in predicting results even one quarter out.

How much more daunting the task for climate-change models, which ideally need to encompass not only the myriad forces on land, in the oceans, and in the atmosphere, about which we have little uncontroversial data and scant understanding, but also an understanding of the Sun and cosmic forces, spanning hundreds of millions of years in the past and many decades into the future. I am not surprised that climate scientists arrive at wildly different conclusions. The suspicious thing would be if they did not. Nor do these disagreements necessarily mean that one group is right and the other wrong, even when they arrive at wildly different results. More likely most of them are right, but only in figuring out one small piece of an impossibly large puzzle.

I do think that many of them are wrong in one very important area, which is how to proceed from scientific results to environmental policy. This applies to scientists on all sides of the global warming debate. Many of them reason along these lines:

"Regardless of whether you believe that nature or man is changing the climate, we should reduce CO_2 emissions. In the worst case, we will have wasted some money but gained a cleaner environment. Implementing Kyoto, or its successor, would be like an insurance policy."

Even some of my deniers hold this belief. On the one hand, they think that climate-change catastrophe is a ludicrous fear. On the other, they think cleaner air or reducing oil dependency would be the upshot of adopting the Kyoto Treaty and aren't too fussed at the waste of money.

But Kyoto is not an insurance policy. Just the opposite, it is the single, greatest threat today to the global environment, because it makes carbon into currency. Carbon is the element upon which all living things are built. With carbon a kind of currency—which is what all carbon taxes and carbon trading and similar schemes do—all ecosystems suddenly have a commercial value that makes them subject to manipulation for gain.

This is not some abstract theoretical concern. We are already seeing environmental havoc from the new economic order that Kyoto has spawned. I know about this havoc mostly from my colleagues at Probe International, who confront it daily. News of it often comes directly from groups in the Third World.

The first big Kyoto calamity is the threat to the world's forests, especially the old-growth forests, which do not soak up carbon from the atmosphere. These have become favorites of corrupt Third World governments. By seizing the forests, cutting them down, and converting them to carbon-intensive plantations, governments and their cronies have been cashing in on carbon credits.

Here's how it was put in the Declaration by the Forest Peoples Program, a gathering of tribal peoples in India's North Eastern Region of Guwahati in 2003: "The climate-change debate has turned forests into a carbon commodity, which will have to provide carbon credits for a lucrative carbon market that will allow industrialized countries to continue emitting greenhouse gases."[1]

Their declaration, which was designed to protest an upcoming Kyoto meeting in Milan, objected to the technocratic mindset that valued their communities on their potential to capture carbon. "The Kyoto Protocol's focus on carbon sequestration means that more credits can be gained the faster a tree can grow, which in turn leads to an incentive for large-scale tree plantations and ignores the role of forests, particularly old-growth forests."[2]

Want another example? Look at the Plantar carbon sequestration project in Brazil's Minas Gerais. It is opposed by more than 50 Brazilian NGOs, citizens' movements, churches, and trade unions, as well as the World Rainforest Movement. The Plantar project—financed by OECD governments and run by the World Bank's Carbon Finance Unit—is converting 23,100 hectares of natural forest to eucalyptus tree plantations to produce wood for charcoal to replace coal for pig iron production. Think of the Plantar project the next time you see an ad suggesting you buy a carbon offset. Projects like these are at the other end of the carbon credit schemes that have become so popular with Westerners. Every time we buy carbon offsets to salve our consciences at flying in a jet, we are helping to dispossess someone, somewhere, by boosting the carbon credit value of their land.

A second looming catastrophe, also caused by Kyoto carbon credits, again affects the poorest of the poor: a dangerous rise in food prices as agricultural lands are turned to ethanol and other biofuels rather than nourishment. In Mexico City last February, some 75,000 people marched in protest at the dramatic rise in the price of tortillas, a corn-based staple of their diet that typically consumes one-third of a poor family's income. Indonesia, Egypt, Algeria, and Nigeria have also seen protests, leading to growing alarm at Kyoto's implications for the Third World. Opinion pieces in the Third World press have been predicting "worldwide famine affecting billions of people"[3] and "political instability, social unrest, and general chaos."[4]

I don't think worldwide famine is likely to materialize, but I do think an increase in hunger is plausible: Goldman Sachs describes the food-price increases as part of a structural change in world agricultural markets, with high-cost marginal lands brought into production to produce fuel. The social and environmental costs will be prohibitive, and not only because wilderness and all manner of ecosystems will be converted to monoculture farming. Ethanol production is water-intensive, requiring 1,700 liters of water per liter of ethanol produced, according to David Pimentel of Cornell University.[5]

To meet this demand, aquifers are being drained and disputes over watercourses are increasing.

Carbon as currency also means the resurrection of large hydroelectric dams and nuclear reactors. Before the Kyoto mindset took hold, these grandiose government-backed relics of yesteryear were struggling to get off the drawing boards of energy planners. With Kyoto's low-carbon chic restoring their respectability, and carbon credits making them less ruinous financially, both are back with a vengeance.

The vengeance could be harsh. The recently completed Three Gorges Dam in China, the world's largest, is proving so dangerous that Chinese officials themselves acknowledge trouble. Fifty-meter-high waves in its reservoir and shoreline collapses of biblical proportions are leading to the relocation of 1 to 2 million Chinese from the Yangtze's banks, in addition to the 1.3 million already moved. The worst-case scenario of a dam failure could lead to a tsunami that would take out downstream Wuhan, a city of 10 million people.

Consider the anxieties raised by Iran's allegedly peaceful nuclear reactor program and then contemplate the risks of a nuclear Armageddon remote if reactors proliferated around the world. Yet many energy planners consider widespread use of nuclear power essential if Kyoto-scale cuts in carbon emissions are to be achieved. Safety systems have failed dramatically in Ontario, Pennsylvania, and other Western jurisdictions that have high engineering

standards, little corruption, and meaningful regulation. Should reactors in large numbers be built in countries where corruption
permits shoddy construction materials, and where regulation is
non-existent, the chance of avoiding future Chernobyls or worse
are poor. Even if reactors don't fail due to a design fault, they are
subject to catastrophe from terrorism. The worst-case scenario of
a nuclear accident is unthinkable—the accident at Chernobyl,
which was far from a major city, cost an estimated $350 billion. A
worst-case accident in North America near a pricey financial center like New York City or Toronto would wipe out much more
than that in property alone, not to mention human suffering.

None of this is to say that we shouldn't be reducing our energy consumption. I'd love to cut back on our energy consumption by 25% or more. We could do this without the need for any
regulation, merely by removing all the overt and hidden subsidies
to road users, industry, and energy producers. This one end-the-
subsidies policy would benefit the economy as it was benefiting
the environment and remove more CO_2 from the air than the
Kyoto Treaty ever could.

I don't expect the subsidies to the energy industry to end any
time soon. With the notion of energy security now a political favorite, the subsidies may well increase, and the environmental
harm too. But I do expect the concern over global warming to
cool. Global warming may be a problem, but it's not a certain
problem, and it's certainly not one of epic proportions, as Al Gore
would have us believe. It is one environmental concern among
many, whose science is far from settled.

Endnotes

Chapter One The Deniers

1 Alex Fryer, "Gore scoffs at Reichert's Stance on Global Warming," *The Seattle Times*, October 25, 2006, http://seattletimes.nwsource.com/html/politics/2003321663_gore25m.html.

2 Sharon Begley, "The Truth About Denial," *Newsweek*, August 13, 2007, http://www.newsweek.com/id/32482.

3 Brian Montopoli, ed., "Scott Pelley and Catherine Herrick on Global Warming Coverage," Public Eye news blog, CBS News, March 23, 2006, http://www.cbsnews.com/blogs/2006/03/22/publiceye/entry1431768.shtml.

4 Ibid.

5 Ibid.

6 Ibid.

Chapter Two The Case of the Disappearing Hockey Stick

1 Daniel L. Albritton et al., *Summary for Policymakers: A Report of Working Group I of the Intergovernmental Panel on Climate Change*, (said document approved in Shanghai when delegations of 99 IPCC member countries participated in the Eighth Session of Working Group I), January 17–20, 2001, 2, http://www.ipcc.ch/pdf/climate-changes-2001/scientific-basis/scientific-spm-en.pdf.

2 Ibid.

3 Stephen McIntyre and Ross McKitrick, "Hockey Sticks, Principal Com-

ponents, and Spurious Significance," *Geophysical Research Letters* 32, no. 20 (2005), http://www.climateaudit.org/pdf/mcintyre.grl.2005.pdf.

4 "Response to Questions Posed by the Honorable Mr. Bart Stupak in Connection with Testimony to the Subcommittee on Oversight and Investigations," July 2006, 7–8, Ans. 6, http://frwebgate.access.gpo.gov/cgibin/getdoc.cgi?dbname=109_house_hearing &docid=f:31362.wais.

List of reviewers of Edward J. Wegman's report before it was sent to committee:

- Professor (emeritus) Enders Robinson, geophysics, Columbia University, elected member of the National Academy of Engineering
- Professor Grace Wahba, statistics, University of Wisconsin-Madison, elected member of the National Academy of Sciences
- Professor Noel Cressie, spatial statistics, Ohio State University
- Professor David Banks, statistics, Duke University, editor, applications and case studies section, *Journal of the American Statistical Association*
- Professor William Wieczorek, geophysics, Buffalo State SUNY
- Dr. Amy Braverman, senior scientist, remote sensing data, data mining, Jet Propulsion Laboratory, Caltech
- Dr. Fritz Scheuren, statistics, NORC, University of Chicago, the one-hundredth president of the American Statistical Association
- In addition, two other reviewers who asked that their names not be revealed because of potential negative consequences for them.

5 Edward J. Wegman, David W. Scott, and Yasmin H. Said, "Ad Hoc Committee Report on the 'Hockey Stick' Global Climate Reconstruction," (report presented to the U.S. House of Representatives Committee on Energy and Commerce), July 14, 2006, 49, http://www.uoguelph.ca/~rmckitri/research/WegmanReport.pdf.

6 Ibid.

7 "Response of Dr. Edward Wegman to Questions Posed by the Honorable Mr. Bart Stupak in Connection with Testimony to the Subcommittee on Oversight and Investigations," July 2006, 9, Ans. 9a, http://frwebgate.access.gpo.gov/cgibin/getdoc.cgi?dbname=109_house_hearing&docid=f:31362.wais.

8 Ibid.

9 "Testimony of Edward J. Wegman before the U.S. House Committee on Energy and Commerce," July 19, 2006, 6–7, http://energycommerce.house.gov/reparchives/108/Hearings/07192006hearing1987/Wegman.pdf.

10 "Response of Dr. Edward Wegman," July 2006, 5, Ans. 5b, http://frwebgate.access.gpo.gov/cgibin/getdoc.cgi?dbname=109_house_hearing&docid=f:31362.wais.

11 "Testimony of Edward J. Wegman before the U.S. House Committee on Energy and Commerce," July 19, 2006, 10, http://energycommerce. house.gov/reparchives/108/Hearings/07192006hearing1987/Wegman.pdf.

12 "Response of Dr. Edward Wegman," July 2006, 5, Ans. 5a, http://frwebgate. access.gpo.gov/cgibin/getdoc.cgi?dbname=109_house_hearing&docid= f:31362.wais.

13 "Testimony of Edward J. Wegman," July 19, 2006, 7, http://energy commerce. house.gov/reparchives/108/Hearings/07192006hearing1987/Wegman. pdf.

14 "Response of Dr. Edward Wegman," July 2006, 2, Ans. 1, http://frwebgate. access.gpo.gov/cgibin/getdoc.cgi?dbname=109_house_hearing&docid= f:31362.wais.

15 Ibid, 3, Ans. 2.

16 "Testimony of Edward J. Wegman before the U.S. House Committee on Energy and Commerce," July 27, 2006, 7–8, http://energycommerce. house.gov/reparchives/108/Hearings/07272006hearing2001/Wegman.pdf.

17 Ibid.

Chapter Three Front Page News

1 *Stern Review: The Economics of Climate Change: Executive Summary*, (The *Review*, which reports to the British prime minister and chancellor, was commissioned by the chancellor in July 2006 and carried out by Sir Nicholas Stern on October 30, 2006.), ii, http://www.hm-treasury. gov.uk/ media/4/3/Executive_Summary.pdf.

2 Richard S. J. Tol, "The Stern Review of the Economics of Climate Change: A Comment," Economic and Social Research Institute, Hamburg, Vrije, and Carnegie Mellon universities, November 2, 2006, 4, http:// www.fnu.zmaw.de/fileadmin/fnu-files/reports/sternreview.pdf.

3 Ibid., 2.

4 *Stern Review: The Economics of Climate Change: Summary of Conclusions*, October 30, 2006, vi, http://www.hm-treasury.gov.uk/media/3/2/Summary_ of_Conclusions.pdf.

5 Richard S. J. Tol, "The Stern Review of the Economics of Climate Change: A Comment," Economic and Social Research Institute, Hamburg, Vrije, and Carnegie Mellon universities, November 2, 2006, 2, http:// www.fnu.zmaw.de/fileadmin/fnu-files/reports/sternreview.pdf.

6 Ibid., 3.

7 Ibid.

8 Ibid.

9 Ibid.

10 Ibid.

11 Ibid., 4.

12 "Experts to Warn Global Warming Likely to Continue Spurring More Outbreaks of Intense Hurricane Activity," media release from Harvard Medical School: Center for Health and the Global Environment, Washington, DC, posted October 21, 2004, http://chge.med.harvard.edu/media/releases/hurricanepress.html.

13 Chris Landsea, e-mail message to Kevin Trenberth and Linda Mearns, October 21, 2004, http://www.nuclear.com/archive/2005/01/20/20050120-001.html.

14 Ibid.

15 "Experts to Warn Global Warming Likely to Continue Spurring More Outbreaks of Intense Hurricane Activity, October 21, 2004, http://chge.med.harvard.edu/media/releases/hurricanepress.html.

16 Ibid.

17 Maggie Fox, "Global Warming Effects Faster than Feared – Experts," Reuters, October 21, 2004, http://www.mindfully.org/Air/2004/Climate-Change-Faster21oct04.htm.

18 Chris Landsea, e-mail message to Dr. R. Pachauri et al. re. "Hurricanes and Global Warming for IPCC," November 5, 2004, http://www.nuclear.com/archive/2005/01/20/20050120-001.html.

19 Ibid.

20 R. K. Pachauri, e-mail message to Chris Landsea re. "Hurricanes and Global Warming for IPCC," November 20, 2004, http://www.nuclear.com/archive/2005/01/20/20050120-001.html.

21 Chris Landsea, e-mail response to Pachauri re. "Hurricanes and Global Warming for IPCC," December 8, 2004, http://www.nuclear.com/archive/2005/01/20/20050120-001.html.

22 Ibid.

23 Chris Landsea," Open Letter to the Community," January 17, 2005, http://sciencepolicy.colorado.edu/prometheus/archives/science_policy_general/000318chris_landsea_leaves.html.

24 IPCC *Fourth Assessment Report, Climate Change 2007: The Physical Science Basis: Summary for Policymakers*, 8, http://ipcc-wg1.ucar.edu/wg1/docs/WG1AR4_SPM_Approved_05Feb.pdf.

25 Pelle Neroth Taylor, "Global Warming Cleared On Ice Shelf Collapse Rap: Natural Causes to Blame, Expert Claims," *The Register*, February 24, 2005, http://www.theregister.co.uk/2005/02/24/ice_shelf_collapsc/.

26 Ibid.

27 Ibid.

28 Ibid.

29 D. J. Wingham, A. Shepherd, A. Muir, and G. J. Marshall, "Mass Balance of the Antarctic Ice Sheet," *Philosophical Transactions of the Royal Society A* 364, no. 1844 (2006): 1627–1635, http://journals.royalsociety.org/content/38315t2244r5w3m4/.

30 Duncan Wingham and Andrew Shepherd, "Ice Thickness Changes in Antarctica and Their Consequences for Global Sea Level," Center for Polar Observation and Modeling, 2006, http://www.cpom.org/research/Fluxes-Ant.htm.

Chapter Four Forging a Consensus

1 R. M. Carter, "The Myth of Human-Caused Climate Change," The AusIMM New Leaders' Conference, Brisbane, QLD, May 2–3, 2007, 65, http://icecap.us/images/uploads/200705-03AusIMMcorrected.pdf.

2 Ibid., 61.

3 Richard S. Lindzen, "Testimony before the U.S. Senate Environment and Public Works Committee," May 2, 2001, 2, http://www.lavoisier. com.au/papers/submissions/Lindzen_McCain.pdf.

4 Ibid., 5.

5 Ibid., 6.

6 Ibid.

7 Richard Lindzen, "Climate of Fear," *The Wall Street Journal*, Opinion Journal Archives, April 12, 2006, http://opinionjournal.com/extra/?id=110008220.

8 Richard S. Lindzen, "Testimony before the U.S. Senate Environment and Public Works Committee," May 2, 2001, 7, http://www.lavoisier. com.au/papers/submissions/Lindzen_McCain.pdf.

9 Ibid., 7.

10 Ibid.

11 Ibid.

12 Richard Lindzen, "Climate of Fear," *The Wall Street Journal*, Opinion Journal Archives, April 12, 2006, http://opinionjournal.com/extra/?id=110008220.

13 Ibid.

14 Ibid.

15 Ibid.

16 R. M. Carter, "The Myth of Human-Caused Climate Change," 65, http://icecap.us/images/uploads/200705-03AusIMMcorrected.pdf.

17 "Climate Change Science: An Analysis of Some Key Questions," Committee on the Science of Climate Change, Division on Life and Earth

Sciences, National Resource Council, 2001, 1, http://www.gcrio.org/OnLnDoc/pdf/ClimateChangeScience.pdf.

18 Richard S. Lindzen, "Don't Believe the Hype," *The Wall Street Journal,* Opinion Journal Archives, July 2, 2006, http://www.opinionjournal.com/extra/?id=110008597.

Chapter Five Is It Warmer?

1 Dr. Vincent Gray, "Support for Call for Review of UN IPCC," (Letter to Prof. David Henderson to support Dr. Henderson's call for a review of the IPCC and its procedures), December 17, 2007, http://nzclimatescience.net/index.php?option=com_content&task=view&id=155&Itemid=1.

2 Ibid.

3 Ibid.

4 R. M. Carter, "The Myth of Human-Caused Climate Change," The AusIMM New Leaders' Conference, Brisbane, QLD, May 2–3, 2007, 66, http://icecap.us/images/uploads/200705-03AusIMMcorrected.pdf.

5 Ibid, 66.

6 Ibid.

7 Ibid.

8 Ibid., 68.

9 Ibid., 66.

10 Syun-Ichi-Akasofu, "Is the Earth Still Recovering from the 'Little Ice Age'?" International Arctic Research Center, University of Alaska Fairbanks, abstract, revised May 7, 2007, 5, http://www.iarc.uaf.edu/highlights/2007/akasofu_3_07/Earth_recovering_from_LIA_R.pdf.

11 Syun-Ichi Akasofu, e-mail and telephone exchanges with author, March 28–29, 2007.

12 Ibid.

13 Ibid.

14 Ibid., 8.

15 Ibid.

16 Ibid., 13.

17 Ibid., 12.

18 Ibid.

19 Ibid.

20 Ibid.

21 Ibid., 13.

22 Ibid., 1.

23 Ibid.

24 A. P. M. Baede et al., "The Climate System: An Overview," *Climate Change*

2001: The Scientific Basis, review ed. B. Bolin and S. Pollonais (2001): 97, http://www.grida.no/climate/ipcc_tar/wg1/pdf/TAR-01.PDF.

Chapter Six Looking for CO2

1 R. M. Carter, "The Myth of Dangerous Human-Caused Climate Change," The AusIMM New Leaders' Conference, Brisbane, QLD, May 2–3, 2007, 64, http://icecap.us/images/uploads/200705-03AusIMMcorrected.pdf.

2 Syun-Ichi-Akasofu, "Is the Earth Still Recovering from the 'Little Ice Age'?" International Arctic Research Center, University of Alaska Fairbanks, abstract, revised May 7, 2007, 2–3, http://www.iarc.uaf.edu/highlights/2007/akasofu_3_07/Earth_recovering_from_LIA_R.pdf.

3 Syun-Ichi-Akasofu, telephone and e-mail exchanges with author, March 28–29, 2007.

4 Ibid.

5 Ibid.

6 Ibid.

7 Ibid.

8 Ibid.

9 Dr. Tom Segalstad, e-mail interview with author, July 5–6, 2007.

10 Ibid.

11 Ibid.

12 Ibid.

13 Ibid.

14 Ibid.

15 Ibid.

16 Ibid.

17 Tom V. Segalstad, "Carbon Cycle Modelling and The Residence Time of Natural and Anthropogenic Atmosphere CO2: On The Construction of the 'Greenhouse Effect Global Warming' Dogma," abstract, Mineralogical-Geological Museum, University of Oslo, Oslo, Norway, 1997, 11, http://folk.uio.no/tomvs/esef/ESEF3VO2.htm.

18 Ibid.

19 Ibid.

20 Dr. Tom Segalstad, e-mail interview with author, July 5–6, 2007.

21 Ibid.

22 Nir J. Shaviv, "Carbon Dioxide or Solar Forcing?" ScienceBits Web Site, undated, http://www.sciencebits.com/CO2orSolar.

23 Ibid.

24 Ibid.

25 Ibid.

26 Ibid.

27 Ibid.

28 Ibid.

29 Ibid.

30 Ibid.

31 Ibid.

32 Ibid.

33 Ibid.

34 Ibid.

35 Ibid.

36 Ibid.

37 Professor R. M. Carter, "Myth of Dangerous Human-Caused Climate Change," Presentation to The Lavoisier Group's 2007 Workshop: "Rehabilitating Carbon Dioxide," Melbourne, Australia, June 29–30, 2007, (said paper was originally published by AusMM in May 2007 and was the paper to which Prof. Carter spoke at the Lavoisier Workshop), www.lavoisier.com.au.papers/Conf2007/Carter2007.pdf.

38 Ibid, 62.

39 Ibid, 62.

40 Ibid, 62.

41 Ibid, 64.

42 Ibid.

43 Ibid.

44 Ibid.

Chapter Seven The Ice-core Man

1 Zbigniew Jaworowski, interviews with author, April 24, 2007–May 5, 2007.

2 Ibid.

3 Ibid.

4 Ibid.

5 Ibid.

6 Ibid.

7 Ibid.

8 Zbigniew Jaworowski, interviews with author, April 24, 2007–May 5, 2007 and December 26, 2007.

9 Zbigniew Jaworowski, "Climate Change: Incorrect Information on Pre-Industrial C02," statement written for the hearing before the U.S. Senate Committee on Commerce, Science, and Transportation, March 19, 2004, http://www.john-daly.com/zjiceco2.htm.

10 Ibid.

11 Ibid.

12 Ibid.

13 Zbigniew Jaworowski, interviews with author, April 24, 2007–May 5, 2007.

14 Ibid.

15 Zbigniew Jaworowski, e-mail correspondence with author, November 24, 2007.

16 Zbigniew Jaworowski, interview with author, May 1, 2007.

17 Zbigniew Jaworowski, e-mail correspondence with author, May 3, 2007.

18 Zbigniew Jaworowski, interviews with author, April 24, 2007–May 5, 2007.

19 Ibid.

20 Ibid.

21 Zbigniew Jaworowski, "Climate Change: Incorrect Information on Pre-Industrial C02," adapted from statement written for the hearing before the U.S. Senate Committee on Commerce, Science, and Transportation, March 19, 2004, http://www.john-daly.com/zjiceco2.htm.

Chapter Eight Models and the Limits of Predictability

1 David H. Bromwich and Thomas R. Parish, eds., "Antarctica: Barometer of Climate Change," Report to the National Science Foundation from the Antarctic Meteorology Workshop in June 1998, published November 1998, 1, http://bprc.osu.edu/gpl/reports/abcc/abcc.pdf.

2 Ibid.

3 "Antarctic Temperatures Disagree With Climate Model Predictions," Ohio State University press release, Ohio State Research Communications Web site, February 15, 2007, http://researchnews.osu.edu/archive/anttemps.htm.

4 Ibid.

5 Ibid.

6 David Bromwich, interview by author, Sept 14, 2007.

7 Ibid.

8 Ibid.

9 Ibid.

10 Ibid.

11 Ibid.

12 Hendrik Tennekes, "No Forecast Is Complete Without A Forecast of Forecast Skill," speech presented at a workshop organized by the European Center for Medium-range Weather Forecasts, April 1986, http://www.sepp.org/Archive/NewSEPP/Climate%20models-Tennekes.htm.

13 Hendrik Tennekes, "The Outlook: Scattered Showers," (based on an October 7, 1986, speech in Riga, Denmark), *American Meteorological Society* 69, no. 4 (April 1988): 1, http://ams.allenpress.com/archive/1520-0477/69/4/pdf/i1520-0477-69-4-368.pdf.

14 Ibid.

15 Hendrik Tennekes, "A Skeptical View of Climate Models," Royal Netherlands Meteorological Institute, http://www.sepp.org/Archive/NewSEPP/Climate%20models-Tennekes.htm.

16 Ibid.

17 Hendrik Tennekes, "Karl Popper and the Accountability of Numerical Weather Forecasting," *Weather* 47 (1992): 343–346, http://climatesci.org/files/popper.pdf.

18 Hendrik Tennekes, "A Skeptical View of Climate Models," Royal Netherlands Meteorological Institute, http://www.sepp.org/Archive/NewSEPP/Climate%20models-Tennekes.htm.

19 Hendrik, Tennekes, "A Personal Call for Modesty, Integrity, and Balance," Climate Science: Roger Pielke Sr. Research Group News, Guest Weblog, January 31, 2007, http://climatesci.org/2007/01/31/a-personal-call-for-modesty-integrity-and-balance-by-henkrik-tennekes/.

20 Ibid.

21 Ibid.

22 Tennekes, "A Skeptical View of Climate Models," http://www.sepp.org/Archive/NewSEPP/Climate%20models-Tennekes.htm.

23 Ibid.

24 Ibid.

25 Freeman Dyson, Winter Commencement Address 2005, University of Michigan, University of Michigan News Service, http://www.umich.edu/news/index.html?DysonWinCom05.

26 Ibid.

27 Ibid.

28 Benny Pieser, "The Scientist As Rebel: An Interview With Freeman Dyson," CCNet, March 14, 2007, http://www.staff.livjm.ac.uk/spsbpeis/Freeman-Dyson.htm.

29 Ibid.

30 Freeman J. Dyson, "The Children's Crusade," in *Disturbing the Universe,* (New York: Basic Books: 2001): 26.

31 Freeman J. Dyson, "The Science and Politics of Climate," *APS News,* May 1999, http://flux.aps.org/meetings/YR99/CENT99/vpr/layub15-01.html.

32 Freeman Dyson, Winter Commencement Address 2005, University of

Michigan, University of Michigan News Service, http://www.umich.edu/news/index.html?DysonWinCom05.

33 Ibid.

34 "Dicastery to Study Global Warming: Experts Will Consider Development and Climate Change," ZENIT Web site: The World Seen From Rome, April 24, 2007, http://www.zenit.org/article-19451?l=english.

35 Freeman J. Dyson, "The Science and Politics of Climate," *APS News*, May 1999, http://flux.aps.org/meetings/YR99/CENT99/vpr/layub15-01.html.

36 "Global Warming Natural, Says Expert: Addresses Vatican Seminar on Climate Change," ZENIT Web site: The World Seen From Rome, April 27, 2007, http://www.zenit.org/article-19481?l=english.

37 Antonino Zichichi, "Meteorology And Climate: Problems And Expectations," [World Federation of Scientists, Beijing, Geneva, Moscow, New York], abstract, presented at Climate Change and Development International Conference, Pontifical Council for Justice and Peace, The Vatican, April 26–27, 2007, 2, http://www.justpax.it/pcgp/dati/2007-05/18-999999/ZICHICHI_METEOROLOGY%20AND%20CLIMATE.pdf.

38 Ibid.

39 Ibid.

40 Antonino Zichichi, "Meteorology And Climate: Problems And Expectations," [World Federation of Scientists, Beijing, Geneva, Moscow, New York], abstract, presented at Climate Change and Development International Conference, Pontifical Council for Justice and Peace, The Vatican, April 26–27, 2007, 0, http://www.justpax.it/pcgp/dati/2007-05/18-999999/ ZICHICHI_METEOROLOGY%20AND%20CLIMATE.pdf.

41 Ibid: 7–8.

42 Ibid. 8.

43 Ibid.

44 Ibid.

45 Ibid., 9.

46 Ibid.

47 Ibid.

48 Ibid.

49 Bob Carter, "Forecasts All Up In the Air," *The Courier-Mail*, June 29, 2007, http://www.news.com.au/couriermail/story/0,23739,21977114-27197,00.html.

50 Ibid.

51 "Climate Feedback: The Climate Change Blog," "Predictions of Climate," entry posted by Oliver Morton on behalf of Kevin E. Trenberth, June 4,

2007, http://blogs.nature.com/climatefeedback/2007/06/predictions_of_climate.html.

52 Ibid.

53 Ibid.

Chapter Nine In the Land of the Midnight Sun

1 Paul E. Damon and Peter Laut, "Pattern of Strange Errors Plagues Solar Activity and Terrestrial Climate Data," *Eos* 85, no. 39 (September 28, 2004): 370, http://stephenschneider.stanford.edu/Publications/PDF_Papers/DamonLaut2004.pdf.

2 Sharon Begley, "The Truth About Denial," *Newsweek*, August 13, 2007, http://www.newsweek.com/id/32482.

3 Graeme Greene, "60 Seconds: Al Gore," (Q&A With Al Gore), September 14, 2006, FridayMetro.co.uk, http://www.metro.co.uk/fame/interviews/article.html?in_article_id=19734&in_page_id=11.

4 Ibid.

5 "New Experiment to Investigate the Effect of Galactic Cosmic Rays on Clouds and Climate," press release from European Organization for Nuclear Research, October 19, 2006, http://press.web.cern.ch/press/PressReleases/Releases2006/PR14.06E.html.

6 Eigil Friis-Christensen, e-mail interview with author, April 11, 2007.

7 Ibid.

8 Ibid.

9 E. Friis-Christensen and K. Lassen, "Length of the Solar Cycle: An Indicator of Solar Activity Closely Associated with Climate," *Science* 254, (1991): 698–700, http://www.sciencemag.org/cgi/content/abstract/254/5032/698?maxtoshow=&HITS=10&hits=10&RESULTFORMAT=&fulltext=solar+cycle&searchid=1&FIRSTINDEX=0&resourcetype=HWCIT.

10 Eigil Friis-Christensen, e-mail message to author, April 8, 2007.

11 Eigil Friis-Christensen, e-mail message to author, April 5, 2007.

12 Ibid.

13 Eigil Friis-Christensen, e-mail message to author, April 8, 2007.

14 K. Lassen and E. Friis-Christensen, reply to article "Solar Cycle Lengths and Climate: A Reference Revisited," *Journal of Geophysical Research* 105 (2000): 27493–27495, http://www.agu.org/pubs/crossref/2000/2000JA900067.shtml.

15 Henrik Svensmark and Friis-Christensen, "Variation of Cosmic Ray Flux and Global Cloud Coverage—A Missing Link in Solar-Climate Relationships," *Journal of Atmospheric and Solar-Terrestrial Physics* 59, no. 11 (1997): 1225–1232.

16 Henrik Svensmark, e-mail message to author, January 4, 2007.

17 Bert Bolin, comment as IPCC chairman quoted in *Information, Copenhagen,* July 19, 1996, (after Svensmark's findings on cosmic rays and clouds were announced at the 1996 COSPAR Birmingham, England, meeting [Original Danish: *Jeg finder dette pars skridt videnskebeligt set yderst naivt og uansvarligt.*]), http://www.apollon.uio.no/vis/art/1998/3/sol.

18 Henrik Svensmark, e-mail message to author, January 4, 2007.

19 Henrik Svensmark, interview by author, January 4, 2007.

20 Henrik Svensmark, e-mail message to author, January 4, 2007.

21 Eigil Friis-Christensen, e-mail message to author, April 8, 2007.

22 Ibid.

Chapter Ten SKY and CLOUD

1 Elizabeth Meen, "Sunspot Activity Hits 1,000-year High," Swissinfo.ch, July 12, 2004, http://www.swissinfo.org/eng/search/Result.html?siteSect=882&ty=st&sid=5080155.

2 Michael Leidig and Roya Nikkhah, "The Truth About Global Warming—It's The Sun That's To Blame," Telegraph.co.uk, July 17, 2004, http://www.telegraph.co.uk/news/main.jhtml?xml=/news/2004/07/18/wsun18.xml&sSheet=/news/2004/07/18/ixnewstop.html.

3 David Schneider, "Living in Sunny Times: Recent Solar Activity Appears to Have Been Unusually Vigorous," *American Scientist* online, January–February 2005, http://www.americanscientist.org/template/AssetDetail/assetid/39261;_f5BEi_EFM.

4 Nir J. Shaviv, "Cosmic Ray Diffusion from the Galactic Spiral Arms, Iron Meteorites, and a Possible Climatic Connection," *Physical Review Letters* 89, no. 5 (July 2002), http://www.phys.huji.ac.il/~shaviv/articles/PRLice.pdf.

5 Nir J. Shavif, "The Milky Way Galaxy's Spiral Arms and Ice-Age Epochs and the Cosmic-Ray Connection," undated, ScienceBits.com, http://www.sciencebits.com/ice-ages.

6 Ibid.

7 Nir J. Shaviv, e-mail interview with author, February 12, 2007.

8 Nir. J. Shaviv, "Cosmic Rays and Climate," undated, ScienceBits.com, http://www.sciencebits.com/CosmicRaysClimate.

9 Ibid.

10 Nir J. Shaviv, e-mail interview with author, February 12, 2007.

11 Henrik Svensmark, "Cosmic Rays and the Biosphere Over 4 Billion Years," *Astron. Nach./AN* 327, no. 9 (2006): 871 DOI 10.1002/ asna. 200610651.

12 Nigel Marsh and Henrik Svensmark, "Cosmic Rays, Clouds, and Climate," *Space Science Reviews* 94 (2000): 215.

13 Ibid.

14 Ibid.

15 Ibid.

16 "Cosmic Rays and Climate," article on the Web site of the Ministry of Foreign Affairs of Denmark, May 10, 2007, http://www.denmark.dk/en/menu/AboutDenmark/ScienceResearch/ResearchAreas/Climate Research/CosmicRaysAndClimate/.

17 The Royal Swedish Academy of Sciences, press office, bio for Jasper Kirkby, Open Lecture at General Meeting for Members of the Academy, February 13, 2008, http://www.kva.se/KVA_Root/eng/events/index. asp?br=ns&ver=6up.

18 "CERN Plans Global-Warming Experiment," physicsworld.com, November 26, 1998, http://physicsworld.com/cws/article/news/3124.

19 "New Experiment to Investigate Cosmic Connection to Clouds," press release from Natural Environment Research Council, October 19, 2006, http://www.nerc.ac.uk/press/releases/2006/cosmicclouds.asp.

20 "New Experiment to Investigate the Effect of Galactic Cosmic Rays on Clouds and Climate," press release from CERN, October 19, 2006, http://press.web.cern.ch/press/PressReleases/Releases2006/PR14.06E.html.

21 Ibid.

22 Scott Capper, "Finding A Heavenly Key To Climate Change," Swissinfo.ch, November 13, 2006, http://www.swissinfo.ch/eng/swissinfo. html?siteSect=105&sid=7253462.

23 Danish National Space Center (DNSC), Web Site, http://www.spacecenter. dk/research/sun-climate/experiments/the-cloud-experiment.

Chapter Eleven Cycles Within Cycles

1 "Odyssey Studies Changing Weather And Climate On Mars," press release 20023-164, NASA Jet Propulsion Laboratory, December 8, 2003, http://marsprogram.jpl.nasa.gov/odyssey/newsroom/pressreleases/20031208a.html.

2 Ibid.

3 Habibullo Abdussamatov, interview via e-mail with author, January 25, 2007.

4 Ibid.

5 Ibid.

6 Ibid.

7 Ibid.

8 Pulkovo Observatory of Russian Academy of Sciences, Pulkovo Web site, http://www.gao.spb.ru/english/astrometr/index1_eng.html; Habibullo Abdussamatov; e-mail interview with author, Jan 15, 2008.

9 Abdussamatov, interview, January 25, 2007.

10 Ibid.

11 Habibullo Abdussamatov, interview via e-mail with author, November 8, 2007.

12 Ibid.

13 Ibid.

14 Ibid.

15 Pulkovo Observatory, http://www.gao.spb.ru/english/astrometr/index1_eng.html.

16 Habibullo Abdussamatov, e-mail interview with author, January 25, 2007.

17 George Kukla, interview via e-mail with author, November 25, 2007.

18 Ibid.

19 Ibid.

20 Ibid.

21 Ibid.

22 George Kukla and Robert Matthews, letter to the White House dated December 3, 1972, in "Origins of a Diagnostics Climate Center," Robert W. Reeves and Daphne Gemmill, authors, http://www.cpc.noaa.gov/products/outreach/proceedings/cdw29_proceedings/Reeves.pdf.

23 George Kukla and Robert Matthews, letter to the White House dated December 3, 1972, in "Global Cooling and the Cold War – And a Chilly Beginning for the United States' Climate Analysis Center?" Robert W. Reeves et al., authors, U.S. National Oceanic and Atmospheric Administration, National Weather Service (2004), http://www.meteohistory.org/2004polling_preprints/docs/abstracts/reeves&etal_abstract.pdf.

24 Ibid.

25 "A United States Climate Program" (1974), cited in "Origins of a Diagnostics Climate Center," Robert W. Reeves and Daphne Gemmill, authors, http://www.cpc.noaa.gov/products/outreach/proceedings/cdw29_proceedings/Reeves.pdf.

26 Ibid.

27 Ibid.

28 Ibid.

29 George Kukla, interview via e-mail with author, November 25, 2007.

30 Ibid.

31 Ibid.

32 Ibid.

33 Rhodes W. Fairbridge, "The 'Solar Jerk,' The King-Hele Cycle, and the Challenge to Climate Science," Climate and Keplerian Planetary Dy-

namics Web site [first published in *21st Century Science and Technology 16*, no. 1: (Spring 2003)], http://www.crawfordperspectives.com/Fairbridge-ClimateandKeplerianPlanetaryDynamics.htm.

34 Richard Mackey, "Rhodes Fairbridge and the Idea that the Solar System Regulates the Earth's Climate," *Journal of Coastal Research* SI 50 [Proceedings of the 9th International Coastal Symposium] (2007): 956, http://www.griffith.edu.au/conference/ics2007/pdf/ICS176.pdf.

Chapter Twelve Better Safe Than Sorry?

1 Dab Elliott, "Ocean Currents to Blame for Warming: Expert," The Sunday Telegraph, April 30, 2007, http://www.news.com.au/dailytelegraph/story/0,22049,21636036-5012769,00.html.

2 Philip J. Klotzbach and William M. Gray with special assistance from William Thorson, "Summary of 2007 Atlantic Tropical Cyclone Activity and Verification of Author's Seasonal and Monthly Forecasts," Department of Atmospheric Science, Colorado State University, November 27, 2007, 45–46, http://typhoon.atmos.colostate.edu/forecasts/2007/nov2007/nov2007.pdf.

3 Cliff Ollier, "The Greenland-Antarctica Melting Problem Does Not Exist," School of Earth and Geographical Sciences, The University of Western Australia, Australia, CCNET, http://www.globalwarming.org/files/Melting%20No%20Problem.pdf.

4 Ibid.

5 Ibid.

6 Ibid.

7 Ibid.

8 "Precautionary Principle," Wingspread Conference on the Precautionary Principle, Science and Environmental Health Network Web Site, January 26, 1998, http://www.sehn.org/wing.html.

9 Ibid.

10 Paul Reiter "From Shakespeare to Defoe: Malaria in England in the Little Ice Age," *Emerging Infectious Diseases* 6, no. 1 (January–February 2000), http://www.cdc.gov/Ncidod/eid/vol6no1/reiter.htm.

11 Ibid.

12 Al Gore, *An Inconvenient Truth: The Planetary Emergency of Global Warming and What We Can Do About It* (New York: Rodale Press, Incorporated, 2007), 172–173.

13 "IPCC Second Assessment Synthesis of Scientific-Technical Information Relevant to Interpreting Article 2 of the UN Framework Convention on Climate Change," 1995, http://www.ipcc.ch/pdf/climate-changes-1995/2nd-assessment-synthesis.pdf.

14 Prof. Paul Reiter, Institut Pasteur, Paris, "The IPCC and Technical Informa-
    tion. Example: Impacts on Human Health," Memorandum to UK House
    of Lords, Select Committee on Economic Affairs, March 31, 2005,
    http://ff.org/centers/csspp/library/co2weekly/2005-09-01/paul.htm.

15 Paul Reiter, interviews via telephone and e-mail with author, March
    20–22, 2007.

16 Ibid.

17 Ibid.

18 Paul Reiter, "Climate Change and Mosquito-Borne Disease," abstract,
    Environmental Health Perspectives 109, S1 (March 2001).

19 Prof. Paul Reiter, Institut Pasteur, Paris, "Statement to Committee on Sen-
    ate Commerce, Science, and Transportation Subcommittee on Global Cli-
    mate Change and Impact," April 26, 2006, http://ff.org/centers/csspp/
    library/co2weekly/20060505/20060505_26.html.

20 Paul Reiter, "Climate Change and Mosquito-Borne Disease," final com-
    ment, *Environmental Health Perspectives* 109, S1 (March 2001).

21 Paul Reiter, "From Shakespeare to Defoe: Malaria in England in the Lit-
    tle Ice Age," *Emerging Infectious Diseases* 6, no. 1 (January–February 2000),
    http://www.cdc.gov/Ncidod/eid/vol6no1/reiter.htm.

22 Ibid.

23 Ibid.

24 Ibid.

25 Ibid.

26 Paul Reiter, "Global-Warming and Vector-borne Disease in Temperate
    Regions and High Altitude," Correspondence: author's reply, *The Lancet*
    351, no. 9105 (March 14, 1998): 839–840.

27 Ibid.

28 Prof. Paul Reiter, Institut Pasteur, Paris, "Malaria in the Debate on Climate
    Change and Mosquito-borne Disease," statement to U.S. Senate Commit-
    tee on Commerce, Science, and Transportation Subcommittee on "Pro-
    jected and Past Effects of Climate Change: A Focus on Marine and
    Terrestrial Systems," "Global Climate Change," April 26, 2006, http://
    commerce.senate.gov/pdf/reiter-042606.pdf.

29 Ibid.

30 Ibid.

31 Ibid.

32 Paul Reiter, "Global-Warming and Vector-borne Disease in Temperate
    Regions and High Altitude," Correspondence, *The Lancet* 351, no. 9105
    (March 14, 1998): 839–840, http://www.sepp.org/Archive/controv/
    controversies/lancetltrs.html.

33 Ibid.

34 Ibid.

35 Ibid.

36 Prof. Paul Reiter, Institut Pasteur, Paris, "The IPCC and Technical Information. Example: Impacts on Human Health," Memorandum to UK House of Lords, Select Committee on Economic Affairs, March 31, 2005, http://ff.org/centers/csspp/library/co2weekly/2005-09-01/paul.htm.

37 Ibid.

38 Ibid.

39 Paul Reiter, interviews via telephone and e-mail with author, March 20–22, 2007.

40 Ibid.

41 Ibid.

42 Ibid.

43 Paul Reiter, "The IPCC and Technical Information. Example: Impacts on Human Health," March 31, 2005, http://ff.org/centers/csspp/library/co2weekly/2005-09-01/paul.htm.

44 Paul Reiter, "From Shakespeare to Defoe," conclusion, *Emerging Infectious Diseases* 6, no. 1 (January–February 2000), http://www.cdc.gov/Ncidod/eid/vol6no1/reiter.htm.

Chapter Thirteen Some Inconvenient Persons

1 "On the Shoulder of Giants," Roger Revelle (1909–1991), Earth Observatory Web site, NASA, http://earthobservatory.nasa.gov/Library/Giants/Revelle/.

2 Melinda Henneberger, "The 2000 Campaign: A Test of Character; On Campus Torn by 60's, Agonizing Over the Path," *The New York Times,* June 21, 2000, http://query.nytimes.com/gst/fullpage.html?res=9806E4D91031F932A15755C0A9669C8B63.

3 Ibid.

4 William Arnold, "Al Gore Takes Urgent Message to the Screen," seattlepi.com (*Seattle Post-Intelligencer*), Thursday, May 25, 2006, http://seattlepi.nwsource.com/movies/271409_goremovie25.html.

5 Ibid.

6 *Omni*, March 1984.

7 S. Fred Singer, "Gore's 'Global Warming Mentor,' In His Own Words," *Environment News*, January 1, 2000, http://www.heartland.org/Article.cfm?artId=9858.

8 Ibid.

9 S. Fred Singer, Roger Revelle, and Chauncey Starr, "What To Do About

Greenhouse Warming: Look Before You Leap," *Cosmos: A Journal of Emerging Issues* 5, no. 2 (Summer 1992): 28–33, http://www.sepp.org/key%20issues/glwarm/cosmos.html.

10 Ibid, 33.

11 S. Fred Singer, "The Revelle-Gore Story: Attempted Political Suppression of Science," in *Politicizing Science: The Alchemy of Policymaking*, ed. Michael Gough, (Stanford: Hoover Press, 2003): 290, http://media.hoover.org/documents/0817939326_283.pdf.

12 Ibid.

13 Ibid., 291.

14 Ibid., 291–293. See also affidavit of Ms. Christa Beran, given August 2, 1993, for Commonwealth of Massachusetts, Middlesex, ss. Superior Court Department, Civil Action, no. 93-2219, cited at Lancaster below.

15 Ibid., 296–297.

16 Ibid., 293.

17 J. Justin Lancaster, "The Cosmos Myth: The Real Truth About the Revelle-Gore Story," (last updated 7/6/06), http://home.att.net/~espi/Cosmos_myth.html.

18 Claude Allègre, *12 Clés pour la Géologie: Les entretiens d'Emile Noël*, (Paris: Belin, 1987).

19 "World Scientists' Warning to Humanity," Union of Concerned Scientists, November 18, 1992, http://homepages.ihug.co.nz/~sai/sciwarn.html.

20 Ibid.

21 Claude Allegre, "The Snows of Mount Kilimanjaro: The Cause of Climate Change Remains Unknown. So, Let Us Be Cautious," *L'Express*, September 21, 2006, http://blog.nam.org/The%20Snows%20of%20Mount%20Kilimanjaro.pdf.

22 Bill Hanley, "The Plowboy Interview" *Mother Earth News*, March/April 1976, http://www.motherearthnews.com/Sustainable-Farming/1976-03-01/The-Plowboy-Interview-Dr-Reid-Bryson.aspx.

23 Reid A. Bryson, Ph.D., D.Sc., D.Engr., "Global Warming? Some Common Sense Thoughts," Global Warming Debunking News and Views Web site, March 20, 2004, http://www.oism.org/news/s49p1837.htm.

24 Bill Hanley, "The Plowboy Interview" *Mother Earth News*, March/April 1976, http://www.motherearthnews.com/Sustainable-Farming/1976-03-01/The-Plowboy-Interview-Dr-Reid-Bryson.aspx.

25 Samara Kalk Derby, "Is Warming Our Fault? Retired UW Prof Claims Humans Aren't To Blame," *The Capital Times*, June 18, 2007, madison.com, http://www.madison.com/archives/read.php?ref=/tct/2007/06/18/0706180285.php.

26 Ibid.

27 Reid. A. Bryson, "Global Warming? Some Common Sense Thoughts," http://www.oism.org/news/s49p1837.htm.

28 Samara Kalk Derby, "Is Warming Our Fault? *The Capital Times*, June 18, 2007, http://www.madison.com/archives/read.php?ref=/tct/2007/06/18/0706180285.php.

29 Prof. David Bellamy, "What A Load of Poppycock!" *Daily Mail*, July 9, 2004, http://ff.org/centers/csspp/library/co2weekly/2005-04-28/load.htm.

30 Ibid.

31 Ibid.

32 Ibid.

33 Ibid.

34 Ibid.

35 Jonathan Leake, ed., "Wildlife Groups Axe Bellamy As Global Warming Heretic," *The Sunday Times*, May 15, 2005, TimesOnline, http://www.timesonline.co.uk/article/0,,2087-1612958,00.html.

36 "Loony IPCC Debunkers Beef Up Security To Deter Protest," Carbon Footprints, ("The Official Blog of Carbon Planet Pty Ltd. Carbon Planet helps you fight climate change by retailing carbon credits to individuals, allowing you to offset your personal CO2 emissions."), http://carbon-planet.com/blog/2007/02/04/loony-ipcc-debunkers-beef-up-security-to-deter-protest/.

Chapter Fourteen The Carbon Catastrophe

1 "Carbon Sinks and Trade, Dams, Rivers Linking and Extractive Industries: New Terms & Mechanisms for Further Expropriation & Livelihoods Threats to Peoples in India's North Eastern Region," Forest Peoples Program Declaration, November 2003, http://www.forestpeoples.org/documents/conservation/carbon_sinks_ips_decl_nov03_eng.shtml.

2 Ibid.

3 Gioetta Kuo, "Food vs. Fuel Wars Just Beginning, *China Daily*, BIZCHINA/Weekly Roundup Website, updated June 7, 2007, http://www.chinadaily.com.cn/bizchina/2007-07/06/content_912170.htm.

4 Ibid.

5 David Pimentel, "Biomass: Agriculture, Ethanol & Biodiesel," in *Encyclopedia of Physical Science and Technology*, (San Diego: Academic Press, 2001): 2.

Acknowledgments

First, I would like to thank those who helped me with the columns that spawned this book: my colleagues at Energy Probe Research Foundation, chiefly Patricia Adams, Thomas Adams, Elizabeth Brubaker, and Norman Rubin, for their thoughtful comments and edits of my drafts; my friend, Richard Owens, whose legal advice provided comfort and perspective when both were needed; and my editor at the *National Post*, Terence Corcoran, for not only allowing a newspaper series with dozens of installments but also encouraging its continuation on the several occasions that I considered wrapping it up.

Most of all, I acknowledge the role in this book of my publisher and editor, Richard Vigilante, and the remarkable crew that he assembled. Richard phoned me one day last fall to suggest I turn my newspaper series into a book. In what seems an instant, it was done. Although my name stands alone as author on the book jacket, this book is as much theirs as it is mine.

Index